NATURE'S MUTINY

NATURE'S MUTINY

How the Little Ice Age of the Long Seventeenth Century Transformed the West and Shaped the Present

PHILIPP BLOM

translated from the German by the author

LIVERIGHT PUBLISHING CORPORATION
A DIVISION OF W. W. NORTON & COMPANY
Independent Publishers Since 1923
NEW YORK   LONDON

The translation of this work was funded by Geisteswissenschaften International –
Translation Funding for Work in the Humanities and Social Sciences from Germany,
a joint initiative of the Fritz Thyssen Foundation, the German Federal Foreign Office,
the collecting society VG WORT and the Börsenverein des Deutschen Buchhandels
(German Publishers & Booksellers Association).

Originally published in German as *Die Welt aus den Angeln* in 2017 by
Carl Hanser Verlag München

For information about permission to reproduce selections from this book,
write to Permissions, Liveright Publishing Corporation, a division of
W. W. Norton & Company, Inc., 500 Fifth Avenue, New York, NY 10110

For information about special discounts for bulk purchases, please contact
W. W. Norton Special Sales at specialsales@wwnorton.com or 800-233-4830

Manufacturing by Lake Book
Book design by JAM Design
Production manager: Anna Oler

Library of Congress Cataloging-in-Publication Data

Names: Blom, Philipp, 1970– author.
Title: Nature's mutiny : how the little Ice Age of the long seventeenth
century transformed the West and shaped the present / Philipp Blom ;
translated from the German by the author.
Other titles: Welt aus den Angeln. English
Description: New York : Liveright Publishing Corporation, a division of
W.W. Norton & Company, [2019] | Originally published in German:
Die Welt aus den Angeln (München : Carl Hanser Verlag, 2017). |
Includes bibliographical references and index.
Identifiers: LCCN 2018052643 | ISBN 9781631494048 (hardcover)
Subjects: LCSH: Climatic changes—Europe—History—17th century. | Climatic
Changes—Social aspects—Europe. | Climatic changes—Economic
Aspects—Europe. | Europe—Civilization—17th century. |
Europe—Intellectual life—17th century. | Glacial climates.
Classification: LCC QC903.2.E85 B5613 2019 | DDC 304.2/509409032—dc23
LC record available at https://lccn.loc.gov/2018052643

Liveright Publishing Corporation
500 Fifth Avenue, New York, N.Y. 10110
www.wwnorton.com

W. W. Norton & Company Ltd.
15 Carlisle Street, London W1D 3BS

2 3 4 5 6 7 8 9 0

Fifth is the race that I call my own and abhor.
O to die, or be later born, or born before!
This is the Race of Iron. Dark is their plight.
Toil and sorrow by day are theirs, and by night
the anguish of death; and the gods afflict them and kill,
though there's yet a trifle of good amid manifold ill.

—HESIOD, *Works and Days*[1]

This is the terrible century about which Scripture speaks
so clearly. It is the Iron Age, which breaks and subdues
all things. The seven angels have emptied their vials over
the earth, and they contained blasphemy, terror, massacres, injustice, treason. . . . We have seen and continue to
see how realm rises against realm, nation against nation,
plagues, famines, earthquakes, terrible floods, signs in
the sun and the moon and the stars; the sufferings of
nations through storms and thunderous waves.

—JEAN-NICOLAS DE PARIVAL, 1654

CONTENTS

ON COMETS AND OTHER CELESTIAL LIGHTS

EPILOGUE: *Supplement to* The Fable of the Bees

The Theology of the Market   272
The Market and the Fortress   277

NATURE'S
MUTINY

Europe where the sun dares scarce appear
For freezing meteors and congealed cold.

CHRISTOPHER MARLOWE,
London, 1578

The lights and windows in the vault of the heavens often grow dark and will no longer shine and shed light on the world's larceny / and they are longing with us for our salvation . . . the sun / the moon / and other stars / shine / less brightly than before / there is no true and lasting sunshine / no steady winter and summer / fruits and things growing from the earth no longer ripen as well / or are as healthy as they most likely used to be.

REVEREND DANIEL SCHALLER, Stendal, in Prussia, 1595

'Twas a harsh cold winter here / you could catch birds and game
    with your hands.
'Twas a hot and arid summer / and the crickets devoured every-
    thing in the field / which caused great price increases.

CHRISTOPH SCHORER, Memmingen, in Germany, 1660

PROLOGUE
Winter Landscape

HOW HAPPY THEY ALL SEEM—moving across the ice with the greatest
of ease. They are either gliding on skates or luxuriously ensconced
in horse-drawn sleighs sailing across the river's polished parquet.
Some huddle together in groups, talking among themselves; others
are playing games. Wealthy gentlemen have their coats slung ele-
gantly across their shoulders, and the ladies are wearing lace caps
or wigs, or both. The simple folk are moving about in short jackets.
There is no fire to warm their freezing limbs, but on this perfect
winter day, no one appears to feel the cold.

The swarming, antlike image of life amid the frost seduces the
eye; the landscape dissolves into a panorama of individual scenes.
The villagers appear in all kinds of situations—from the two lovers
in the haystack (are they both men?) to the naked behind poking out
of a broken boat and a second bottom whose owner is half hidden
by a willow tree; from the mother with her child in the foreground
to the men playing golf, the reed cutter with his enormous load, and
the young couple gliding hand in hand across the hard surface. A
woman drinking from a beaker is one of the few figures whose face
is revealed. Most of the villagers are moving away on their wooden
skates with thin steel blades—toward the horizon, into a future that
is little more than a sketch.

Slightly to the right of center stands an important-looking
group in elegant, gold-embroidered garb—ladies with hoop skirts
and tall wigs, gentlemen with precious ostrich feathers adorning

their hats. A gray beggar attempts to stir their pity, but they show no interest. What are they doing on the ice in some godforsaken village—without a coach, without servants? How did they get here? And what exactly are all these people doing? They are not celebrating any special event or feast day, not Christmas and not Carnival. It is not even Sunday—the church looming in the background is eerily dark.

The longer one examines this panorama, the less plausible it becomes. The initial impression of realism grows into an understanding of allegory. A whole society is on the same ice here—rich and poor, men and women, children and the elderly, masters and servants—all equalized by frost and cold but seemingly untouched by it. Only the animal cadaver in the left corner hints that death will leave its mark on this idyll; the bird trap made from an old door is ready to snap shut and crush its next victim, reminding the viewer that earthly pleasures are transient—*Carpe diem* (Seize the day). An empty beehive stirs latent memories of lazy summer afternoons with blazing, colorful blossoms. Soaring above this busy miniature world, exactly in the middle of the sky, a large bird seems to fly higher and higher. Is it an ordinary bird or the last memory of the presence of the Holy Ghost?

The creator of this landscape, Hendrick Avercamp (1589–1634), specialized in winter scenes. He painted them year-round in his workshops in Amsterdam and later in Kampen, in the Netherlands. Depicting bustling crowds socializing while apparently not feeling the cold may have been an expression of his own longing to be part of a greater world. He was born deaf and mute, lived quietly with his mother, and followed her into the grave within a few months of her death. The happy scenes he painted were always the lives of others.

Like all painters of his time, Avercamp created his compositions not from direct observation but from sketches and memory. Artists worked in their studios with paints that were freshly prepared, and it would have been a major technical challenge to transfer an entire studio outside. According to the ideas of the time, it would have seemed meaningless simply to paint what happened to be there. Not nature itself, but its symbolic meaning was of greater interest to artists and their patrons. A landscape was a skillful composition of disparate elements, not merely the documentation of reality. Aver-

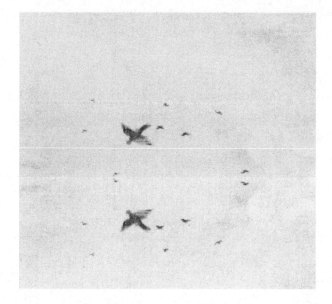

Detail, birds fly overhead.

camp's sketches were assembled into a whole, an image of the social world in a deceptively natural setting—another reason why his landscapes consist of many individual scenes.

Avercamp and his colleagues composed and constructed communities on the ice, creating an underlying sense that we must all share the same resources, we are all exposed to the same circumstances—a message characteristic of the Netherlands, where collective and cooperative work had been long established among farmers for maintaining dykes, draining marshes, and managing fishing. The aristocracy had never been strong here, and the convictions that everyone shared the same stretch of land and that people swim or sink together had deep roots in cultural practices and attitudes.

The people on the ice were assembled there by Avercamp's imagination. The evocation of the season, however, was entirely realistic. Winter landscapes had only recently become a fashionable subject for artists, who sold their paintings mainly to wealthy urbanites. Avercamp created his little masterwork in 1608, a year that had suffered particularly severe frosts. Half a century earlier, winter landscapes had been almost unknown in European painting. Since then, however, winters had been unusually long and grim, snow fell more abundantly than before, rivers and seaports remained frozen until the spring. Word had it that in Eastern Europe, birds dropped out of the sky, frozen to death.

The winter of 1607–8 was one of the severest in recorded history—and not only in the Netherlands, where the rivers and canals had become icy panoramas on which painters could imagine entire societies. In London, the Thames was frozen so solid that a Frost Fair was erected on it, a semipermanent setup consisting of kiosks and wooden huts with taverns and even brothels. Henri IV of France awoke one morning to find his beard iced over; wine froze solid in its barrels, and deep snow covered parts of Spain. Europe was a frosted world.

The fierce cold spells had begun in the 1560s and 1570s, and artists responded quickly to what must have been a dramatic change in nature. Pieter Bruegel's *Hunters in the Snow*, painted in 1565 after a particularly severe winter, is a vivid evocation of the inhospitable harshness of the season. This magnificent work also marks a change in painting style. The *Hunters* was painted as part of a cycle of the seasons, an artistic convention that reached back into the Middle Ages. There are little winter landscapes in the famous *Très riches heures du duc de Berry* (1412–1416), but always as part of a seasonal cycle, indicating that spring would always arrive after the bitter cold of the dark winter. Now, however, winter landscapes established themselves as a genre in their own right, and this even invaded other themes. Still following the medieval tradition of painting biblical stories in familiar surroundings, the Adoration of the Magi and the Massacre of the Innocents—both episodes from the New Testament—were now set in deep snow.

Avercamp's landscapes describe this frigid world and hint at the new social order that would emerge from it. Everyone on the ice is sharing the same cold. On the wide expanse of the frozen canal, they even resemble each other. They all have to step carefully, or they will fall. The elegant gentlemen and the poor eel fisherman with his long trident, the sleigh with its proud white horse and the gaggle of peasants—they are all challenged to find ways of coping in this unfamiliar and inhospitable environment.

Almost a century later, in 1691, the English composer Henry Purcell would distill the European experience of a hundred years of cold into the famous, frozen, halting lament of the "Cold Genius":

> What power art thou, who from below
> Hast made me rise unwillingly and slow
> From beds of everlasting snow?
> See'st thou not how stiff and wondrous old
> Far unfit to bear the bitter cold,
> I can scarcely move or draw my breath?
> Let me, let me freeze again to death.[1]

≋

A SIMPLE QUESTION appears at the beginning of this book, a question with an undeniably contemporary dimension: Do societies change when the climate changes? And if so, how? Which immediate and secondary effects flow from a change in the natural framework of societies, of cultures and mental horizons? The long, wintry seventeenth century serves as a test case for investigating this question and mapping the effects of climate change on all aspects of life, from agriculture to philosophy.

Even in a historical context, it is true that the science of climate change is clouded by controversy. When did the Little Ice Age begin? What caused it? How violent were its effects, and what long-term consequences did it have? Some climate historians argue that its beginnings were in the Middle Ages and that it lasted until the early nineteenth century; others concentrate on the period of greatest deviation from the statistical mean, lasting roughly from 1570 to the 1680s. I will concentrate on this second interpretation, because it raises other interesting issues.

The period of particularly disturbed climatic patterns and events that stretches from the late sixteenth to the late seventeenth century also was an age of tremendous social, economic, and philosophical upheaval in Europe. At its beginning, we see a feudal world centered on castles and churches; at its end, we encounter a world of cities and markets, of nascent capitalism and the vigorous early stirrings of the Enlightenment. Can we assume that these natural and cultural transformations could be related? If so, how?

The Little Ice Age was a global phenomenon that led to a reduction of average temperatures by about two degrees Celsius, though with radically different effects on local levels. Ocean currents and the salinity of seawater were affected, oceanic condensation patterns changed, polar ice caps and glaciers grew rapidly, climatic systems were disturbed. All this led to a succession of brutal and extreme weather events—severe storms, weeks of rain, and even years of summer drought, as well as unrelenting frosts. There were reports of population collapse due to famine in Ming-period China, as well as murderous winters in North America and paltry harvests in India, while the Ottoman Empire experienced some of the most devastating winters its historians had ever recorded.

Three reasons lead me to limit my focus to Europe. First, the Little Ice Age has been particularly well researched for Europe, and there is a great wealth of scientific and documentary source material yielding a very detailed picture of how people experienced the changes and how they adapted to them. Second, I lack the expertise and language skills to delve into the cultural histories of Japan, China, India, Malaysia, and the Aztec Empire and to relate cultural practices and their changes to climatic and other factors. I would be entirely dependent on translations and interpretations of others. Third, the period of the Little Ice Age in Europe marked particularly momentous changes in its societies. Thus, while it is important to stress the global dimension of climate change in the seventeenth century, I will focus on the European experience.

⊰⊱

THE CULTURAL HISTORY of climate has, of course, its own long trajectory. Aristotle and Hippocrates already had written about the connections between climate and culture, and during the seventeenth century, the historian Montesquieu formulated thoughts on the interrelationship between natural environments and the societies in them. During the twentieth century, climate also made its most spectacular appearance front and center in the French *Annales* school, whose expanded view of social history incorporated living conditions and their changes.

Most influential for Western philosophy, however, was the comprehensive philosophical geography penned by Georg Wilhelm Friedrich Hegel, a German philosopher who left his native country only once, for a short spell as a private tutor in Bern, Switzerland. Hegel argued that the spirit of a culture resembles the landscape and the climate in which it grows, which is why he felt so confident that only the German landscape with its forests and its temperate climate was suited to being "the true scene of world history," because it created true spiritual depth. He also insisted that indigenous Americans and Africans were unable to build a great culture because "cold and heat are here too powerful to allow a mind to construct a world for itself."

At the turn of the twentieth century, pseudoscientific racial thinking in the West was comfortable with the idea that other, less-developed cultures had emerged from their natural circumstances, while the colonial powers themselves had transcended the bounds of nature and reached a higher plane of civilization. It was not until after World War II that historians began to look at the idea of climate (though not yet climate change) as a determining factor for all civilizations. Fernand Braudel, with his studies of societies and trading connections in the Mediterranean, and Emmanuel Le Roy Ladurie, who reconstructed life in medieval southwestern France, made clear that instead of speculating about the effects of climate on culture, as Hegel had done, a much more fruitful method was to use historical data to create finely grained analyses of local circumstances.

Today, historians such as Geoffrey Parker, Jared Diamond, and Christian Pfister have greatly added to our knowledge and understanding of this issue. They argue that climatic factors were instrumental in the rise and fall of entire civilizations. The perspective on the meteorological past of the planet and all its varied cultures has widened and deepened, despite the fact that most research to date has focused narrowly on climate change as a meteorological phenomenon, with less emphasis on its cultural consequences.

Mass migrations and the decline of great civilizations, in particular, have aroused a good deal of scientific interest: The end of the Roman Empire is frequently linked to a period of unusually cool temperatures in the mid-fourth century CE, which was caused by a volcanic eruption whose ash created a global winter for some years, affecting not only ancient Rome but also far-distant cultures from China to Peru. But these incidents occurred far in the past; documentation is rudimentary, especially with regard to understanding the effects of climatic changes on the societies of the day. The Little Ice Age, on the other hand, with its vast range of sources, documents, and data, provides an ideal case study of the subtle interactions between climate change and cultural change.

⋇

SCIENTISTS ARE NOT yet able to agree on the causes of the global cooling that manifested itself in the brutal winters and cold summers of the 1560s onward. Although there is abundant evidence of what went on, there is very little understanding of exactly why it occurred. For the purposes of this book, which concentrates not on the causes but on the consequences of the Little Ice Age, a brief summary of the current state of research may suffice.

The earth is a vast archive recording its own history. Climate historians and climate scientists can come to a very precise, highly localized understanding of what happened year by year via several methods: analyzing cores drilled out of the polar ice shelves or out of glaciers; measuring the distance between tree rings as an indication of stressful or easy growing seasons; and identifying plant residues such as pollen, spores, and leaves in layers of mud or other sediments to find out what characteristic flora and fauna existed in a particular period. In this way, scientists have created detailed climate maps for different eras and areas.

In addition to the availability of these scientific data, Europe also offers a surprising wealth of human data and historical documents. Diaries and letters, weather observations and sermons, literary works, wine-harvest dates, the packing lists and logbooks of merchant seamen, paintings and account ledgers show not only the immediate effects of climate change on the natural world but also what transformations they may have caused or accelerated in social and cultural contexts.

This mosaic can be assembled into a portrait, though that portrait may vary according to which interpretation of the data one chooses to follow. From around 1400 (and for reasons not yet clearly understood), global temperatures were depressed by around one degree Celsius, but the decline was more severe from the second half of the sixteenth century onward. Temperatures in Eurasia and particularly the Atlantic region dropped more sharply, by around two degrees Celsius.

This change was accompanied by extreme weather events. There was increased seismic activity, with earthquakes and volcanic eruptions intensifying, possibly because the growing polar ice shelves made seawater more saline, thus changing the temperature and speed of deep-sea currents, which in turn exerted different patterns of pressure on continental shelves and their oceanic borders. This, however, is merely a hypothesis. That multiple large volcanic eruptions hurled vast amounts of dust into the atmosphere and added to the harshness of winters by blocking out sunlight, however, is a fact.

This very condensed account raises many questions, first and foremost that of causality. Why did this global cooling occur? No one knows. The most promising candidate is a change in solar activity, but the evidence of this change (a decreased number of sunspots recorded by early observers) dates only from decades after the beginning of the Little Ice Age. Was human activity to blame? Not this time. One theory posits that the European invasion of South America caused the local populations to collapse due to imported viruses such as smallpox (the native South Americans reciprocated by giving Europeans syphilis) and therefore caused agriculture to retreat, allowing rain forests to reclaim once-farmed areas and thus increase the absorption of CO_2. But this effect would account for only a very small amount of the dramatic temperature drop—no more than one-tenth of a degree—especially since at the same time European forests were being leveled dramatically, evening out the global balance.

While causalities and contributing factors remain unclear, the effects of this change are richly documented. The first wave of bitter winters, rainy summers, and violent hail and late frosts hit Europe during the second half of the sixteenth century, and this was particularly catastrophic for agriculture, devastating harvests everywhere. A two-degree drop in annual mean temperature translates into almost three weeks of lost growing time, meaning that crops were very much slower to ripen, and sometimes failed to ripen at all.

In the first instance, this caused a long-term, continent-wide agricultural crisis. Failed harvests, wheat rotting in the field or being destroyed by hail, and long periods of drought in other European areas meant that, after 1570, the amount of wheat harvested would not reach pre-1570 levels for 180 years, until 1750.

In our very different world, removed as most of us are from agricultural processes, it requires effort to appreciate what this meant for people at the time. Sixteenth-century Europe relied primarily not on trade and industry but on local agriculture, particularly cereals—wheat, barley, rye, and oats—as well as wine or beer (more grain), and, in Mediterranean regions, olives. Fresh fruits and vegetables were available only in season or if they had been preserved; maritime fish (with the notable exception of pickled herring) were available only in coastal regions. Meat was too expensive for most Europeans to eat regularly, and the hunting of wild game was a privilege jealously guarded by the aristocracy.

Thus, for their daily diet and ultimately their survival, most people were dependent mainly on grain—in bread, soups, porridge, or dry biscuits—with the occasional addition of animal proteins and some fresh or preserved fruits and vegetables, depending on the season. Any disruption in grain production had potentially devastating consequences. Hunger would be widespread, leaving people more vulnerable to epidemics and sometimes causing them to riot, or, if they could, simply to pack up and move away. Marginal agricultural regions such as Northern and Eastern Europe—where, even in normal climatic conditions, temperatures had barely been high enough to grow staple crops—were hit particularly harshly.

The relatively sudden cooling of temperatures tore into a social and economic system that had been working relatively well for almost a millennium—a system centered on feudal landowership and on close cooperation between Crown and Church. Peasants (and serfs in many European regions) toiled just as generations had done before them. Many had no metal plows, and their wooden tools would barely scratch the surface of the soil. Only the wealthier farmers had oxen to pull their plows, since it was costly to keep ani-

A well-deserved rest: Harvest was the pivotal point of the agricultural year.

mals fed through the winter. Many peasants simply had to do their
own work, putting themselves or other family members to the har-
ness. Most areas operated on the principle of rotational cultivation,
allowing one of every four fields to lie fallow for a year, so the soil
could recover. This was particularly necessary since much of European agriculture was effectively a near-monoculture, thus draining the soil of important elements and over time leaving fields less
productive. Yields were low: One grain sown would result in only
about four grains harvested, a fraction of what would be achieved
in later eras.

During the seventeenth and eighteenth centuries, European
agriculture would diversify. Columbus's expeditions to America had
already brought such plants as maize, potatoes, and tomatoes back
to the "old continent," but they mainly graced botanical gardens
and private collections. Potatoes, for instance, were popular with
botanists because their blossoms were beautiful, but it took a long
time for rural folk to accept these filling tubers as food. The vast

majority of European agriculture still consisted of grain production, and most of it was consumed locally.

The all-important harvest would be split into three parts: one for the family and the farm animals, one for seed to be kept for sowing in the spring, and one (often the rest of what remained) for the lord, as tax. This part would be determined by the size of the harvest: In good years, the peasants had to pay much more in tax, which also meant that they had no incentive to be productive. Any increase in bushels harvested or sacks of grain filled was simply destined to vanish into the lord's own granaries, from which much might then be sold on to the towns and cities.

As harvest failures became more frequent, this ancient system began to falter. Peasants watched helplessly as their crops failed to ripen, or were destroyed by rain, frost, thunderstorms, or hail. This meant, of course, that the peasants went hungry, but it also meant that they could no longer pay their taxes. For the aristocracy, relying for their income almost totally on their lands, this meant potential ruin. Typically barred from working in the trades or the professions, the aristocrats' primary economic duty was to provide resources for warfare: not only food stocks but also horses, munitions, manpower, or simply cash. In addition, aristocrats needed to maintain impressive living standards and demonstrate generosity to their followers, if they were to retain their social standing and their power in the land. Aristocrats always needed money. A hungry, rebellious population, unable or unwilling to pay their taxes, might be dealt with on a sporadic basis, but when occasional shortages became an ongoing crisis, the entire social order was threatened.

While the peasants (who formed the greater part of the European population) lived in the countryside, and the aristocracy typically lived alongside them, those in the towns and cities were also threatened by failing harvests. The single most important commodity in any city was grain, and flour prices (and hence bread prices) served as something of a gold standard at the time. Controlling and manipulating grain prices was one of the very few policy tools available to a late medieval ruler or city council. The price of bread could mean the difference between hunger and plenty, between social peace and rampant crime, between public calm and rioting in the streets. As more harvests remained below expectations and less grain was available, bread prices rose rapidly, at times doubling or trebling within a year, taking the daily bread, as it were, from the mouths of increasing numbers of city dwellers.

LIFE WITHOUT MONEY

In the countryside, traditional structures existed for coping with poverty and maintaining relatively stable rural communities. In England, France, Germany, and throughout Central Europe, for instance, village communities usually had access to a common— a piece of land on which everyone, even the landless poor, could graze a goat or a few chickens and harvest animal feed for the winter. Life in the countryside remained very different from life in the city. Most villagers and most peasants operated at subsistence level and would have only occasionally come into contact with actual money. They could make or barter for most things they needed, and their few other necessities could be purchased on market days, which occurred at regular intervals, often around religious feast days. These provided opportunities to sell a ham and a few eggs, to buy a length of cloth or a handful of nails, and even to look for a wife, or simply to have a good time.

Even in the cities, the social and economic life was unlike what we would recognize today. There was a developing world of trade, but the amount of wares reaching all of Europe from Asia and from other continents within any given year would have fitted into a single modern containership. Long-distance trade was concerned almost exclusively with luxuries such as spices, porcelain, silk, and tobacco.

The typical city dweller's outlook on life and business would also be unfamiliar to modern eyes. Not only were bread prices tightly controlled, but so were the prices for other commodities. Crafts and manufacturing (such as it was) were in the hands of guilds, which could determine not only the standards to which a master had to work and train his apprentices, but also how many masters could have shops of their own, what they could produce and how, and what they could charge for their goods.

This controlled urban world valued social capital—class and family standing, trustworthiness, cooperation—but it did not encourage anyone to reach beyond his station. A tailor who did well for himself might think of buying another house, donating a new window for his church, supplying a new ornament for the guildhall, or even subsidizing the pay for a dozen soldiers—and he would make his donation known more or less discreetly among the people of the town. He was unlikely, however, to deposit money in a bank to accrue interest, or to think of opening branches of his shop in other cities, thus increasing production and pursuing the goal of rising through the social ranks. You were born a tailor or had been apprenticed to one, and that meant that your highest ambition in life would most likely be to be a good tailor, respected among your townspeople, accruing civic honor, if you could, rather than pots of gold.

If this sounds idyllic, it is mostly so only in hindsight. Most European societies were encased on all levels in an ironclad economic protectionism and corporatism, with each person expected to remain in his allotted station, and with an idea of human inequality deeply rooted in social perceptions and social practice. This stifled most private initiative, enterprise, and innovation, and it clearly divided the vast majority of people into different strata. Though the Church provided mobility to a handful of the brightest village boys, if you were born a peasant, it was overwhelmingly likely that you would die a peasant, and so would your children.

This inflexible social world was nonetheless not completely stable, buffeted as it was not only by famine and epidemics but also by warfare, with armies plundering crops or even simply destroying them as they passed through. The consequent loss of life was frequently appalling, and those who remained alive in a devastated region often had no recourse but to migrate. By the end of the sixteenth century, for instance, in France, the Netherlands, Flanders, Germany, Central Europe, Hungary, and Italy, religious wars lasting three whole generations had left hundreds of thousands dead, maimed, or orphaned. The succession of severe winters and sunless summers that began around 1570 was one more bitter turn of the screw for the continent's already hard-pressed populations.

THE GREAT EXPERIMENT

What happened next, during the long seventeenth century, seems almost something ordained by the callous, testing god of Job, or by some extraterrestrial scientist conducting an experiment with an entire biological system, including the varied populations and societies of *Homo sapiens*. What happens if one changes a system's parameters—the temperature, the weather, the climate? What will collapse and what will endure? Who will live and who will die? Will those creatures whose very existence is threatened find some way to escape? Will they then find some way, despite all that has changed, to establish themselves again, and to flourish?

This idea of a distant demiurge suggests the bird's-eye perspective imagined by the Dutchman Hendrick Avercamp and his painter colleagues, who in the early seventeenth century produced exquisite winter landscapes peopled by countless tiny, bustling figures. There is value, as well as danger, in this perspective. It reveals patterns that cannot be seen from the ground, but it also invites generalizations where they are not necessarily legitimate. It represents the temptation to tell a grand story at the expense of providing the details. Perhaps the details are interesting, but they complicate things with a more accurate portrayal of the often-contradictory nature of lived reality.

Any history of transformative processes must acknowledge their fundamentally untidy and frequently paradoxical nature. The Little

Ice Age, too, has a complex texture of developments and counter-currents, of asynchrony between societies and cultures, and between urban and rural areas; different stages of social or economic development can coexist. The city folk of the seventeenth century lived in a world that was, from our perspective, a good century ahead of the agrarian realm of some of their kin.

There were also broader geographical divisions, so that similar developments sometimes occurred centuries apart. Italy, for example, was ahead of the rest of Europe not only in its exceptional cultural flowering in the fourteenth century but also in developing sophisticated banking and financial services that would not be introduced in Russia until five hundred years later. But the Little Ice Age also saw a dramatic decline in Italy's leading role, leaving others to take the lead politically as well as culturally. Mighty Habsburg Spain, on whose global empire the sun never set, was also subject to serious reversals and a loss of power and influence, while the Netherlands—until then home to herring fishermen, farmers, and a handful of town merchants—suddenly surged and became the greatest naval power of its day, as well as a center for economic, artistic, and even philosophical renewal.

This story, then, does not describe one unified march in a single direction, but rather many meandering routes at a time of serious and unpredictable change. Faced with new factors of climate, some societies reacted and others did not; some reacted wisely and others foolishly. Medieval and comparatively modern ways of thinking and living existed simultaneously in this period of contradictions and asynchronicities. Modern and antiquated weapons were used by the same armies and in the same wars; the first stirrings of scientific theory coexisted with religious mysticism; new ideas lived alongside, and sometimes confronted, old beliefs.

Unlike the simulations we are accustomed to when we speak of climate change today, this real-life experiment was not conducted under laboratory conditions. Outcomes were partly determined by preexisting cultural and economic factors, beyond conscious control. Other outcomes were forced or hastened by the innovative, the daring, and the ruthless: Luther and Leonardo, Columbus and Gutenberg. Great movements, springing from many sources, transformed Europe profoundly: the Renaissance, the Reformation, the discoveries of other continents and other creatures (significantly, not mentioned in the Judeo-Christian Bible). As with other cultural innovations, their influence was much stronger in the cities than in the countryside. The Renaissance remained largely an elite cultural—and to some extent political—phenomenon; the religious Reformation transformed some areas of Europe while leaving others relatively untouched; printed broadsides, pamphlets, and books were interesting only to those who could read, although stories about strange happenings in faraway places were, as they had always been, part of everyday folklore. Climate change, however, affected everyone. There was no escaping the weather.

The agricultural crisis caused by the Little Ice Age served as a catalyst for change everywhere, facilitating some ideas and practices—social, cultural, and political—while making others more difficult, or even, in the long run, impossible. Existing feudal structures groaned and cracked and sometimes split apart entirely.

In this book, we will follow the different paths people took in their search for a way out of this agricultural impasse—some paths unsuccessful, some succeeding spectacularly. Beginning with a late-sixteenth-century world where theologians had the last word, where science and medicine followed the allegorical models of antiquity, and politics focused on the church and the fortress as mental and physical landmarks, where everyone was dependent on local grain production and the local powers-that-be, where the Earth was held to be the center of the universe and the sun was believed to revolve around it. After a thousand years of stasis, quite suddenly, within only four generations, this world gave way to one in which we today, centuries later, can readily recognize ourselves—where people began to talk of markets, empirical knowledge, and human rights, and where reason and pragmatism moved increasingly to the forefront of European cultural, social, and political life.

We must be on our guard, all the same, with this linear understanding of history, for it can also obscure an important truth: Though most of these changes occurred in reaction to a particular set of circumstances, they usually occurred without deliberate planning, and all of them occurred without any certain foreknowledge of outcomes. They were born from immediate needs, from experiment, from intuition, and also from fear and greed. Those that failed have left little trace; others, inconclusive, are no more than footnotes to history. Those changes that did establish themselves, however, continue to an often surprising extent to provide the framework for our lives today, and to what we can imagine.

The first part of this book depicts the beginning of the Little Ice Age in Europe around 1570; it sketches the immediate human consequences of this crisis and reveals how witnesses saw and explained it. How did they react to the cold spells and the storms? What did they say about them, and what did they do? What did they think was causing them?

It was clear to many observers around the beginning of the seventeenth century that their societies had entered a time of crisis, even if they were not yet certain what role nature played in it and what role belonged to God. Though the crisis may have begun in agriculture, its wide ramifications required a much broader response than simply addressing agricultural concerns. Consequently, the second part of this book examines the economic, social, scientific, military, and cultural developments that occurred during this period, and their mosaic-like correlations with environmental change.

The third part of this book goes a step farther and follows the changes that were triggered by the Little Ice Age and its agricultural crisis on European thinking. Climatic upheavals in the seventeenth century encouraged the empirical observation of natural processes, helping to develop broader mental horizons and bolder attitudes that would bring, chainlike, ever more change to societies not just in Europe but around the globe. We usually call this way of thinking the Enlightenment.

An epilogue gathers the book's thematic strands and links them with the climatic, political, and cultural changes occurring in the present day. We have inherited so much more from the seventeenth century than may be obvious at first glance. One idea in particular, formulated for the first time in Europe around 1600, was to allow the continent to gain a position of spectacular global dominance: The medieval acceptance of human economic life as cyclical and stable was rejected in favor of the idea of continuing economic growth based on exploitation. This was to prove the generator of European wealth, built on relentless imperial and industrial expansion. It is this same idea of growth based on exploitation that now poses so clear a threat to the well-being of our species.

✂

DURING THE SEVENTEENTH CENTURY, a small army of Flemish and Netherlandish painters produced canvas after canvas depicting idyllic winter landscapes of frozen rivers, little villages, bare forests, and great expanses of white—the ideal stage for a social panorama of people living and laughing in the cold. These paintings sold for handsome sums, mainly to comfortable burghers inhabit-

The hunters' return: As the cold was closing in, artists responded
by creating a new genre, the winter landscape.

ing the proud houses of Antwerp, Bruges, and Amsterdam. These represented a new class of patrons, interested not in great displays of power and prestige but in simpler yet subtler dramatizations of Christian virtues, set in a world they knew and understood. The little figures careening across the icy width of these canvases seem not to have a care in the world. But they are fictions. Their real-life counterparts lived in fear and increasing uncertainty.

"GOD HAS ABANDONED US"
Europe, 1570–1600

God shows us his anger,
by sending us eternal winter,
cold which we feel at home,
wrapped in our thickest furs.

CONTE MARCO ANTONIO MARTINENGO,
Brescia, Italy, 18 May 1590

A MONK ON THE RUN

Wouter Jacobszoon did not want to witness history in the making. He had chosen a life of quiet contemplation and had become abbot of a monastery in Gouda. Much of the Netherlands was still Catholic then, but loyalties wavered in the long-festering war with the Spanish occupiers, and Wouter's town eventually threw in its lot with the Protestant rebels, the *geuzen*. For a monk, this was a dramatic event, as Catholics were suddenly reviled, and were hunted and even murdered by vengeful mobs and hardened rebels. In 1572, like thousands of others, Abbot Wouter Jacobszoon made his way from Gouda to still-Catholic Amsterdam.

The abbot was no longer a young man. He was about fifty years old when he put pen to paper to record what was happening to him and his brethren. Even in Amsterdam, the situation for them was precarious. There were severe tensions between Calvinists and

Catholic loyalists, but as a monk Wouter was able to live discreetly and keep out of trouble. His former life of responsibility and autonomy, however, was gone. In his exile, all he could do was to wait for peace, and to pray. In the meantime, he would entrust to his diary his fears and hopes, his observations and the news he gleaned while walking around in the streets. At some point in the future, he must have hoped, people would understand what he and his fellow Catholics had endured for their faith.

Time and again, Wouter implored the Lord for help, pleading and attempting to convince Him that He had been silent for too long, that it was time to intervene on behalf of the faithful, who were suffering so grievously at the hand of heretical rebels. While he was waiting for some sign from God, Wouter jotted down rumors of an impending peace, as well as news of executions, massacres, tragic personal stories and moments of hope, and even the prices of butter and grain. He also wrote about the exceptional cold that worsened the suffering of the orphaned, the homeless, and the hungry. The Lord was testing the country severely. Sometimes Wouter was overcome with despair. "God has abandoned us, the grace of the Holy Sacraments was taken from us,"[1] he noted in the year of his flight.

Throughout the early pages of Jacobszoon's diary are a lot of fervently muttered prayers. "O heavy, o oppressive, o crushing times. Who could not quake, and who not shudder?" he wrote in September 1572. "Between hope and fear we are driven on and yet we only go from day to day. Jesus Lord, Jesus Son of David, Jesus of Nazareth, finally reveal your divine power and do not forget us in the hour of our most dire need."[2] Soon, however, the entries become more pragmatic, almost breathless in their urgency. Events are moving at an overwhelming pace. Wouter himself is suffering from "hunger and grief."

> During this time, the weather was bitter cold. Everything froze and became hard. It hailed, it snowed, and cutting winds were whistling from All Souls Day [November 2] to now [in March].

The good Lord, we learn from this, wants to show us thus how much we have gone astray, but the people did not change and behaved as if they were his enemies. Like wolves and lions they attacked not only men of the cloth but also simple folk, good country people and everyone they found.[3]

People had to fend for themselves as best they could, and the diarist kept records of their often desperate attempts to survive. In early November 1572, a group of farmers attempted to salvage the cadavers of drowned cows in order to eat them. The *geuzen* had requisitioned them and driven the entire herd across the ice, but the animals had broken through the too-thin layer. Considering that it

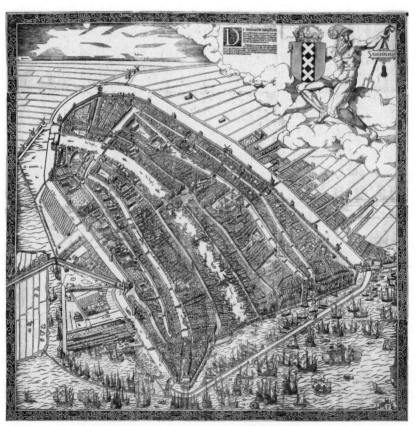

Safe harbour: Amsterdam became a refuge for many
Catholic refugees during the religious wars.

was early November, what is remarkable is not that the ice was too thin to support a herd of cows, but that there was any ice at all. The miscalculation of the rebels now caused desperate country folk to venture out onto the ice in order to try to feed their families.

War-plus-cold is an appalling combination. In the rebel areas, grain, herring, and other staple foods became increasingly unaffordable—if they were to be found at all outside the black market. Every meal seemed a piece of special good fortune. Even the troops defending the city found themselves fighting not mainly against the enemy but against hunger and also sickness, ever-present in time of war and exacerbated by malnutrition. "The poverty, which was suffered in this time, is beyond description . . . and it frequently happened that ten or twelve soldiers a day were carried from the inn to the graveyard and were buried without coffins, in simple mats."[4] It was becoming more and more difficult to defend the starving city.

The snow stayed until April, "as if it was still winter," and when Jacobszoon finally ventured out of Amsterdam in July to visit a priest friend in Haarlem, he could hardly recognize his own country:

> On the way I saw the terrible destruction which has been visited upon us in this time of confusion. I saw very few houses between Haarlem and Amsterdam which were not burned down. All churches we saw along the way were either totally burned out or badly damaged or destroyed. In many places the country was deserted and bare of animals [in the meadows]. I also saw . . . a naked person lying in the middle of the street, right in the track of the carts, dried out and crushed by the wheels, so that a decent person would be shocked to see this. And it was strange that nobody was to be found to clear away this corpse and to cover it with soil, so that it simply lay there, left to the wild animals.[5]

In his Amsterdam asylum, the monk continued to believe in the possibility of peace. Many people, however, had given up hope. Almost every page of the diary mentions children dying of hunger and women committing suicide—with continuing high winds and endless, driving rains exacerbating the general misery. A storm in November 1574 caused a flood "so that one could take a boat from one house to the other," after a dyke had broken. When winter set in, the flooded fields became icy deserts. In January, a woman was found sitting on the street between Amsterdam and Haarlem. Still holding her baby to her breast, she had died of cold and was already frozen solid. The child, however, was alive. Spanish soldiers took it with them to their camp.

><

WOUTER JACOBSZOON'S DIARY is one of very few sustained, personal accounts about life at the beginning of the Little Ice Age, describing the day-to-day experience of living with ever more severe weather conditions. Many observers, however, had already concluded that something unusual, something dark and threatening was taking place. During some years, the sun shone only dimly, even at the height of summer, before ceding to interminable, ice-locked winters. The exceptional frosts particularly captured the public imagination. In 1569, the lagoon of Venice had stayed frozen until March, and during the winter of 1572–73, when Wouter Jacobszoon found refuge from the rebels but not from the cold in Amsterdam, Lake Constance, between Germany and Switzerland, was covered with thick ice until well into the spring.

In Stendal, Prussia, the Protestant pastor Daniel Schaller was also keeping a daily record. During the 1590s, he noted:

> The lights and windows in the vault of the heavens often grow dark and will no longer shine and shed light on the world's larceny / and they are longing with us for our salvation. And as the windows in an old house grow dark / and the face on a body that has lived too much / so the old and cold world also / declines visibly / the sun / the moon / and other stars / shine / less brightly than before / there is no true and lasting sunshine / no steady winter and summer / fruits and things growing from

the earth no longer ripen as well / or are as healthy as they most likely used to be.[6]

Nature itself seemed to be turning against humanity:

> Nowadays, at the end of the world, one expensive year follows another without cease / and there is not only a great scarcity of bread, and very dear prices for our beloved corn and grain / but everything / from the smallest to the largest / which is needed for keeping house and for sustaining life / has come to the dearest penny / and has risen in the extreme. . . .
>
> The field and the soil are tired of bearing fruit and are exhausted / and in the towns and villages one hears / much wailing and lamenting about this / for it brings further price increases and famine. . . . The wood in the forests no longer grows / as it did in the olden days . . . therefore *ruina mundi* [the ruin of the world] must be at our gates.[7]

Schaller's religiously inspired pessimism may sound extreme, but modern dendrochronologists—scientists who analyze tree rings and gather information about plant growth and climate—agree with the pastor. In the cooler conditions of that era, trees really were growing more slowly, causing the prices for wood to rise as well, just when cities needed more firewood to keep their houses warm through the winter. Among those few who kept diaries and recorded their observations of this harrowing time were several men of the cloth, especially Protestants, since they were more likely to be literate and to have the time to keep regular notes. Schaller's colleague Thomas Rörer, from Giengen an der Brenz in southern Germany, about 350 miles (almost 600 kilometers) from Prussian Stendal, described the effects of the great cooling on the wine producers in his area: "Even the soil itself seems to decline. The vineyards no longer want to give so much good wine, the fields not so many bushels of wheat, and the trees no longer such good fruits, as some years ago."[8]

Schaller feared "revolt, rebellion and unrest," noting that no fewer than ten earthquakes had shaken Prussia since 1510, a clear portent for the theologian, who sensed "certainly the signs of the Last Judgment and the last quake in which / all dead wake up / and come out of their graves / before Christ's judgment."

GOD'S WIND AND WAVES

Despite the men of God preaching the End of Days, the Last Judgment did not arrive as they predicted. But the natural world remained in the grip of fierce change, and not only on land. In 1588, the British historian William Camden recorded the delight of English fishermen who discovered that shoals of herring, normally found on the high seas hundreds of miles to the north, were now appearing off the coast of England.

The oceans were cooling. In pursuit of their prey, sperm whales were found beached on the shallow coasts of the North Sea, and storms from the Arctic buffeted ships in European waters. It was one of these storms that caused the greatest maritime disaster—or, depending on which side one takes, the greatest military miracle—of the sixteenth century.

Elizabethan England was at war with Spain, and in August 1588 a gigantic Spanish fleet set sail due north from the Basque harbor of La Coruña, planning to board an auxiliary invasion force in the Spanish Netherlands and from there to descend upon the island of the Virgin Queen. The Armada, as the fleet was called, the pride of the Spanish navy, was the greatest fleet assembled since the days of antiquity. Some 130 vessels crowded the sea: men-of-war, converted merchant ships, armed galleys from Naples, and smaller craft, with a combined firing power of twenty-five hundred cannon, manned by eight thousand seamen and an invasion force of eighteen thousand soldiers, to be supplemented with a further three thousand waiting to board in the Netherlands—a terrifying and seemingly invincible army.

Armada, the armed one. The world's mightiest fleet
could not withstand unseasonal arctic storms.

The plan had been months in the making. King Philip II and his commanders had overseen a vast logistical operation and had planned the invasion with great care. Even so, for the Spanish, it ended in disaster. One salient detail the planners had failed to take into account was the shallowness of the waters along the Dutch coast. Most of the Spanish ships had deep drafts that precluded them from approaching the coast to pick up their waiting soldiers. Dutch ships were flat-bottomed and agile in the lively coastal winds, and thus highly effective in disrupting any craft sailing between the coast and the ponderous, heavily armed ships that lay farther out at sea.

The commander of the Spanish fleet, Admiral Don Alonso Pérez de Guzmán y de Zúñiga-Sotomayor, duke of Medina Sidonia, refused to contemplate a retreat now with a view to making a second attempt sometime in the future. It was late summer already, and the window of opportunity for a successful invasion of England was small. The duke determined to sail on without the extra soldiers, but misfortune continued to dog his expeditionary force. Fighting with English defenders in the Channel weakened the invaders. Once anchored in English waters, the Spanish fleet was attacked by English fireships sent into their midst, which forced the Spanish to cut their anchors and retreat, to prevent their vessels from being engulfed in flames. Facing stubborn opposition from the English, harassed by Dutch rebels against Spanish rule, and with many of their ships already damaged, the Spaniards' invasion already had effectively failed.

But the tragedy had only just begun. Not daring to risk further fighting in the Channel, with his forces diminished and many of his ships in need of repair, Admiral Guzmán decided instead to sail around the British Isles and thence return to Spain. Even so, part of his fleet was barely saved from sinking. The refitted merchant-men, not built to support the weight and recoil of heavy cannon, had been structurally shaken in the fighting, and several had to be tied together with ropes around their hulls just to keep them from falling apart. Hoping to reach Spain before the weather turned, the Armada was forced to sail north toward Norway and from there down the west coast of Scotland and out beyond Ireland, a very long detour on the journey home to Spain.

But many of the men aboard these Spanish ships were never to see their homeland again. In September, when the fleet reached the open sea west of the Orkney Islands, it was caught in a severe storm; the ships were pressed onto the rocky coast of western Ireland. Having jettisoned their anchors off the English coast, many of them were unable to take refuge in the relative safety of a bay, where they might have waited until the punishing icy gales had passed and the roaring sea was becalmed.

The Spanish were excellent sailors, lords of a global empire and used to navigating in different conditions and climate zones. But even they could not have foreseen a storm of this size, completely atypical as it was in those latitudes. Reconstructions of climate data and the logbooks of the captains indicate that this was a hurricane, of a different magnitude entirely than the normal September winds off the Irish coast. It was, in effect, an Arctic storm, possibly intensified by the rising temperature differences between warm air in the atmosphere and the expanding Arctic ice shelf, which had already rendered northern Greenland inaccessible by ship for months every year.

The Spanish lost twenty-four vessels of their great Armada, sunk or pressed onto the rocks of the Irish coast. Five thousand crew and soldiers were killed, either by the storm itself or by pitiless Irish wreckers and plunderers, who felt it safer to leave no Spanish survivors to tell their story. Of the 130 ships that had set out from La Coruña, only sixty-seven returned, and all were badly damaged. On board, many thousands of men lay desperately sick and close to starvation. The fanatically pious King Philip II, who had been determined to re-Catholicize Britain by force, was stunned: "I sent my ships against men," he is reported to have lamented, "not against God's wind and waves."

HARSH FROSTS AND BURNING SUN

An Arctic storm off the Irish coast in September had saved Queen Elizabeth's Protestant realm from a Catholic invasion. Less welcome aspects of climate change, however, were also making themselves felt in England. Between 1400 and 1550, London's great and normally temperate artery, the River Thames, had frozen over five times (in 1408, 1435, 1506, 1514, and 1537); between 1551 and 1700, it was covered by a thick layer of ice no fewer than twelve times: in 1565, 1595, 1608 (the year Avercamp was painting his village on the ice in nearby Flanders), 1621, 1635, 1649, 1655, 1663, 1666, 1677, 1684, and 1695.

One of these perishingly cold years, 1666, illustrates another characteristic of climate change during the period. Not only did temperatures fall, but they also became generally more extreme and less predictable; the icy winter might be followed by an extremely dry and warm spring, and a particularly hot, arid summer. In the silence of the London night, the beams of thousands of timbered

In 1666, the great fire of London destroyed a large part of the city. A particularly hot and dry summer had preceded the fire.

buildings groaned and cracked as they lost the last residues of their usual moisture, turning tinder dry. It took nothing more than a small fire in a bakery in Pudding Lane on September 2 to transform the English capital into a sea of flames that burned for three days and devoured some thirteen thousand houses. When this Great Fire of London was finally spent, the old city had been reduced to ashes and some eighty thousand Londoners had lost their homes.

The baneful influence of the weather produced strong societal effects. Violent protests increased as grain prices rose. There is a clear correlation between years with extreme weather and riots and rebellions, and this is especially pronounced in years when the harvest was poor. When the years of the Little Ice Age are compared with those of the previous two centuries, the pattern is easy to see: There were some twenty riots due to high grain prices in the British Isles between 1347 and 1550. Toward the end of the sixteenth century, however, incidents of this kind began to multiply, receiving increasing public attention, with violence and unrest in affected areas often stretching over long periods. In the years from 1585 to 1660 alone, more than seventy such uprisings are documented.[9]

The bitter weather has left its mark in the literature of the time. It was during the icy season of 1595 that William Shakespeare wrote his great history *Richard III*, with the villainous protagonist famously opening the play with the line "Now is the winter of our discontent. . . ." And in *Coriolanus*, the tragedy of a Roman army leader cast out by his own people because of his excessive pride, Shakespeare begins with the Roman poor rioting over the price of bread. In 1608, as he was working on this very text, bread riots gripped the streets of London and the Thames was covered in a thick coat of ice.

Many Elizabethan poets integrated into their verses the experience of living through severe winters. In his 1612 drama *The White Devil*, John Webster wrote of "Cold Russian winters, that appear so barren, / As if that nature had forgot the spring." In the same work, Webster reflects on a longer historical development nearer to home: Southern England had possessed vineyards since Roman times. During the fourteenth century, English winegrowers had even exported their wares to France—much to the annoyance of French vineyard owners. By the seventeenth century, however, the cold had put an end to English wine production. Webster's writing reflects the fact that even the most dedicated care could not save the precious plants:

> As in cold countries husbandmen plant vines,
> And with warm blood manure them; even so
> One summer she will bear unsavoury fruit,
> And ere next spring wither both branch and root.

Christopher Marlowe had found even stronger words for the recent European experience in his *Tamburlaine the Great*, written during the wet and cold year of 1587: "Europe where the sun dares scarce appear / For freezing meteors and congealed cold." Francis Bacon—lord chancellor, scientist, essayist, and universal genius—

was one of the first European intellectuals to record his fascinated concern at the recent and dangerous changes in the weather. He was also one of the first to realize that it was not a local phenomenon, but had much wider implications:

> They say it is observed in the Low Countries (I know not in what part), that every five and thirty years the same kind and suit of years and weather comes about again; as great frosts, great wet, great droughts, warm winters, summers with little heat, and the like; and they call it the prime. It is a thing I do the rather mention, because, computing backwards, I have found some concurrence.[10]

Like the inhabitants of every great city at every time, seventeenth-century Londoners were always looking for new entertainments and the chance to make a few extra pennies. The congealed cold of the frozen Thames now became a playing field for the population. Frost Fairs were held on the ice, visited by thousands eager to see the different attractions in dozens of huts and tents, including open fires on which whole oxen were roasted on spits. Canny printers took their presses onto the ice and produced souvenir flyers in front of the shivering onlookers.

But the harsh conditions also drove countless people from their homes, refugees from warfare, from religious persecution, or, if they were living in areas where agriculture was marginal and difficult to begin with, from the endless gray skies and ruined harvests. In the play *Thomas More*, collaboratively written with other playwrights, Shakespeare contributed a monologue in which More addresses a crowd rioting against refugees. It is the only dramatic text to survive in a manuscript copy by Shakespeare's own hand:

> Imagine that you see the wretched strangers,
> Their babies at their backs and their poor luggage,
> Plodding to the ports and coasts for transportation,
> And that you sit as kings in your desires,

Authority quite silent by your brawl,
And you in ruff of your opinions clothed;
... say now the King [would] banish you: whither would you go?
... Why, you must needs be strangers, would you be pleas'd
To find a nation of such barbarous temper
That breaking out in hideous violence
Would not afford you an abode on earth.
Whet their detested knives against your throats,
Spurn you like dogs, and like as if that God
Owed not nor made not you, not that the elements
Were not all appropriate to your comforts,
But charter'd unto them? What would you think
To be us'd thus? This is the strangers' case
And this your mountainish inhumanity.[11]

Shakespeare made Thomas More an advocate for a common humanity at a time when war, famines, and freezing temperatures were cutting into the lives of millions. France was hit particularly badly by harsh winters and bad harvests at a time when the country was riven by religious wars and the deprivation attendant on them. Between 1562 and 1598 alone, some four million French men and women became victims of war, civil war, starvation, and epidemics.[12] Extreme climate events exacerbated the suffering of the population. In 1570, rivers as far south as Provence and the Languedoc froze over. In 1594, the harbor in Marseille became unnavigable because of ice; the frost finally appeared to loosen its grip in February, only to return with renewed intensity in April.

Drought and continuous rain during the summers were particularly severe threats to agriculture, and as supplies grew less predictable, hunger itself became a weapon of war. In the spring of 1590, during his bloody dispute with the Catholic League, the Protestant king of France, Henri IV, determined to retake Paris, his capital, by siege. He could not secure sufficient artillery for a sustained bombardment and a decisive attack, so he decided to starve the city.

On May 7, his army surrounded Paris, burned all windmills in the vicinity, and blocked the streets through which food usually reached the city's markets.

The people of Paris had only just survived a particularly harsh winter. Now, in the early summer, a pitiless sun burned down on them. Armed mobs demanded peace or bread, but the city's defenders refused to yield, though their situation was becoming more desperate by the day. Hunger swelled the stomachs of children and adults, and the streets were lined with victims too weak to move, or those already dead. After the horses and donkeys had been eaten, cats and dogs were brought to communal kitchens, where a broth of their meat was prepared in large cauldrons. Many people chose the most desperate individual escape, drowning themselves in the Seine.

Only the Jesuits and the Capuchin monks appeared to have enough biscuits, salted meat, and vegetables in their stores to feed themselves for another year—as the king's secretary found to his astonishment. Not until the city administration forced their hand did these men of the Church agree to share their hoard with the hungry. Even so, by June, the situation was so desperate that the Spanish ambassador, Dom Bernard de Mendoza, suggested to the astonished relief committee that the bones of the dead at Paris's famed *Cimetière des Innocents* should be crushed and ground to a flour to be baked into bread. By the time the city was relieved by Catholic troops on August 30, 1590, some forty-five thousand people, one-fifth of the population of Paris, had starved to death or succumbed to deadly illness.

A TIME OF CONFUSION AND A FIERY MOUNTAIN

Farther to the east, the famous black earth of the Ukraine and the Baltic states had made this region of vast fields and small populations the breadbasket of Europe. Large quantities of grain had long been shipped across the Baltic Sea toward Amsterdam and from there as far as southern Italy, yet now those producing the grain

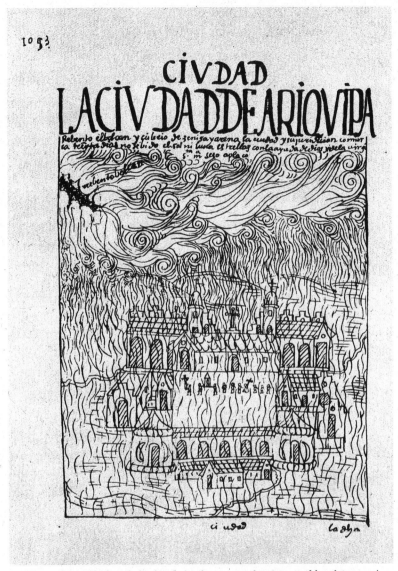

A mountain explodes: Ash clouds in the atmosphere caused by the eruption of Mount Huaynaputina in Peru caused famines in faraway Russia.

rarely had enough to feed their own families. A single bad harvest could have devastating results, especially since many landowners found it more lucrative to sell grain at a profit than to leave enough of it to their serfs—feudalism would long remain a prevalent form of social organization in Eastern Europe. In 1601, after a cold and rainy summer during which the harvest, almost ripe, had rotted in the sodden fields, famine broke out in Russia. More bad harvests followed, and rebellions and civil war made the situation ever more desperate as rival powers fought for the throne of Russia. It is estimated that between 1601 and 1603 some three million people in Russia and neighboring areas succumbed to famine and fatal illness. Over the course of those two years, the disputed Russian Tsar Boris Godunov lost perhaps a third of his subjects.

The famine only deepened a political crisis known in Russia as *Smutna*, the time of troubles or confusion. Following the fatal stabbing of Dimitri, son of Ivan the Terrible, the throne was occupied by Boris Godunov, who may or may not have arranged the murder himself. In 1601, however, a man appeared in Poland claiming to be the real Dimitri, heir to the throne, still alive after all. This charismatic claimant, one of several messianic figures of the time, soon gathered thousands of followers. In the ensuing civil war, the so-called false Dimitri even succeeded in occupying the throne, but in 1606, he, too, was murdered. Russia was rent by waves of violence which continued spasmodically until 1613, leaving the land exhausted and the cities desperately in need of reconstruction—a task not undertaken, however, until the end of the seventeenth century, when Tsar Peter the Great began his vast, ruthless campaign of modernization at all costs.

One of the prime causes of Russia's manifold troubles around the turn of the seventeenth century was in fact found thousands of miles away. It was documented by the Peruvian scholar Felipe Guáman Poma de Ayala, quite unaware of the far-distant consequences of the dramatic event he had observed. One folio of his early seventeenth-century chronicle, *El primer nueva Corónica y buen gobierno*, shows the Peruvian town of Arequipa covered in a dramatic rain of ash. Poma de Ayala commented: "The volcano exploded and covered the town and the administrative area with ash and sand. For thirty days one could see neither sun, nor moon, nor stars. With the help of God and the holy virgin Maria it stopped, and calmed down."[13]

Poma de Ayala had witnessed the eruption of Mount Huayna-putina in Peru on February 19, 1600. With the first tremors and clouds of dust thrown into the sky, indigenous tribes had begun preparing to make assuaging sacrifices, including of young girls, to their gods, with the Spanish Catholic priests in the region attempting to do the same with their own rituals: processions, prayers, and exorcisms. It was all to no avail.

The volcano erupted during an Indian ceremony of sacrifice, hurling thirty cubic kilometers (more than seven cubic miles) of ash and rocks up to twenty miles (thirty-five kilometers) into the stratosphere. The rain of ash fell over an area of many hundreds of miles, and the flash floods caused by the eruption traveled seventy-two miles (120 kilometers) to the Pacific coast. But the eruption also had a major effect on the weather in Europe. As ice-core analyses have shown, the sulfuric acid that entered the atmosphere with the volcanic ash triggered a volcanic winter that caused years of cooler temperatures across the Northern Hemisphere.

The eruption of Mount Huaynaputina once again raises a question of causality. Was the Little Ice Age a consequence of increased volcanic activity, as is sometimes claimed? In all likelihood, the chain of events runs in the opposite direction: Volcanoes added to the effect of climate change but did not cause it. It has been argued, however, that an increase in salinity of ocean waters may have changed the temperature and the course and speed of deep-sea currents, thus also altering the pressure they exerted on the continental shelves. This changing pattern of pressure on geological formations and continental drift may in turn have contributed to the increase in volcanic eruptions and earthquakes reported during this period. This, however, is merely a hypothesis. The climatic system is immensely complex, and ultimately, though our modern analysis is a vast improvement on the understanding of these issues that people may have had during the seventeenth century, it is still not great enough to allow full comprehension of the complex, multidimensional system that is our global climate.

PILGRIMS AND THEIR HUNGER

In North America, too, global cooling was a mortal threat. There is little evidence of how Native American tribes suffered and survived during this period, but the first European settlers left fascinating insights into their lives, and their deaths, during this period. The first colonists of Jamestown in today's Virginia found blossoming landscapes when they made landfall in the spring of 1607. The 104 arrivals were greeted by "faire meddowes and goodly tall Trees, with such Fresh-waters running through the woods, as I was almost ravished at the first sight thereof," as George Percy, one of the settlers, would later write.[14]

Nature was so abundant with animals, fruits, and wonderful farming soils that the settlers, established in a walled camp, were happy enough to be left behind as Captain Newport, one of the officers responsible for the expedition, sailed back to England; and this despite the fact that, according to Percy, the settlers were "verie bare and scantie of victualls, furthermore in warres and in danger of the Savages." Surrounded by the earthly paradise they had found, a land of fabulous natural wealth and seemingly endless resources, the settlers felt confident they would survive and prosper.

Later, however, Percy's report records a marked change of fortune, as hostile natives and hostile nature began to tighten the noose, and the expected rains failed to arrive. "The sixt of August there died John Asbie of the bloudie Flixe. The ninth day died George Flowre of the swelling. The tenth day died William Bruster Gentleman, of a wound given by the Savages, and was buried the eleventh day."[15] The litany of deaths continued, while stocks of food began to run out. More serious, however, was the constant siege by the local Indians. And most threateningly of all, the grain the colonists had sowed failed as arid heat strafed the land for weeks on end, and even the wild animals retreated into the forests.

> Thus we lived for the space of five moneths in this miserable distresse, not having five able men to man our Bulwarkes upon

any occasion. . . . If there were any conscience in men, it would make their harts to bleed to heare the pitifull murmurings and out-cries of our sick men without reliefe, every night and day, for the space of sixe weekes, some departing out of the World, many times three or foure in a night; in the morning, their bodies trailed out of their Cabines like Dogges to be buried.

Weakened by hunger, the colonists had little resistance to an increasingly inhospitable environment. Deaths from disease were a daily occurrence in the small community, but, as Percy wrote, "Our men were destroyed with cruell diseases . . . but for the most part they died of meere famine."[16] Without relief and without food, some colonists took to exhuming the corpses of those who had died, in order to eat them. Some were close to madness. One man murdered his wife and secretly preserved her body in salt. Another, Hugh Pryse, blasphemously declared in the marketplace that he, if he were God, would not countenance such cruelty. Pryse and another man decided to leave the walled camp and were soon killed by Indian warriors.

When a supply ship finally reached them toward the end of the year, only forty colonists were still alive. They could not know, as we now know from modern climate analysis, that their once-paradisiacal region had just experienced its driest summer in 770 years.

TRUTH AND WINE

The effects of the Little Ice Age in Western Europe were less devastating than in Russia, China, or North America, but they were still significant enough to cause considerable economic and social disruption. Owing to the lack of reliable and comprehensive data, it is difficult to estimate the rise and fall of European grain harvests, but the situation is very different with regard to wine. Wine was a precious commodity, so from the sixteenth century onward, it was meticulously documented, including wine descriptions, prices, harvest amounts, and, most important, harvest dates.

Ripe grapes on the vine are prey to birds and thieves, to rot and mold and frost. Winegrowers therefore try to harvest as early as possible in order to obtain the maximum quantity of grapes, but at the same time, they need to wait as long as possible to ensure the grapes have had enough time to ripen, thus ensuring the highest possible quality. Before the fruit becomes overripe, every sunny day creates more sugar in the grape, more and more delicate aromas. Consequently, the harvest date must be chosen with great care, and the harvest must be completed quickly. In cooler years, however, the fruit does not ripen sufficiently. In such situations, the vintners can either wait and run the risk of losing much of their harvest to birds and rot, or they can harvest unripe grapes with too little sugar and too much acidity, which will result in sour wine. Harvest dates and wine descriptions can therefore give a surprisingly accurate picture of the weather.

Until well into the nineteenth century, drinking wine was much more than a luxury, especially in the cities, where it was often impossible to find clean drinking water. Since the germ-killing virtues of boiling water had not yet been discovered, the only solution lay in adding alcohol to drinking water—i.e., drinking watered-down wine or beer instead of plain dirty water. As harvests became more difficult and poor supplies meant that wine of poor quality was sold at ever-increasing prices, beer established itself as a more secure alternative in many Northern European countries. While also reliant on the weather, wheat, hops, and malt were simply less sensitive and less vulnerable to weather fluctuations than grapes were.

The habit of enriching drinking water with alcohol also meant that our city-dwelling ancestors—men, women, and children— were seldom completely sober. Servants, laborers, journeymen, soldiers, and even many prisoners had a right to a certain amount of wine each day—and more for those who were working hardest. A hike in wine prices created almost as much of a social problem as inflated bread prices, and commentators noted every variation in the weather as crops were repeatedly damaged, decimated, or simply unripe at harvest time. The Zurich theologian Heinrich Bullinger

Wine harvest under their lordships' watchful eyes: Wine
was an everyday part of the diet of many Europeans.

noted in 1570: "The spring of this year was like winter, cold and
wet, the wine blossom terrible, and the harvest bad."[17] Eighteen
years later, the Cologne patrician Hermann Weinsberg noted with
grim humor that in spite of his name (literally: vineyard), he had
drunk his last wine—his cellar was empty. For thirteen years, he

had been unable to find wine worthy of aging, "because the vineyard now freezes and no longer produces wine."[18]

The most comprehensive documentation of wine dates and
prices comes from Burgundy, whose protected climate allowed delicate wines to ripen more reliably than elsewhere. During the Little
Ice Age, the region was able to profit from this geological protection against weather extremes, but even so, the effect of cooler years
was evident. While the average harvest date between 1500 and 1570
occurred around September 27, it was two days later from 1571 to
1600, and the quality of this late-harvest wine was deemed inferior
to earlier vintages. In the less-protected Jura, however, just a little
farther to the east, the difference between harvest dates in the two
periods mentioned was more than a week, while the wines in the
Alpes Maritimes were harvested almost two weeks later.

In the climatically more vulnerable region around Paris, the vines
felt the effects of a drop in average temperatures even more. Before
1570, the harvest had started roughly on September 21; between 1571
and 1620, the date was ten days later: October 1. In the particularly
cold year of 1573, just as Wouter Jacobszoon was describing the misery of life in the war-torn Netherlands, the wine harvest began on
October 16; in 1581 and 1587, it didn't occur until the end of that
month. The data clearly show how capricious nature had become,
resulting in disturbed or changed weather patterns with alternating
rainy seasons and hot, arid periods. In 1599, the wine harvest was held
during the first week in September; one year later, it was in the second
week of October, a difference of some six weeks.[19]

WINE IN VIENNA

About six hundred miles to the east of Burgundy, in Vienna, the
situation was remarkably similar. The city's largest wine producer
was the Bürgerspital, the burghers' hospital, with about a hundred
hectares of vineyards, whose produce not only kept patients happy
but also was sold to cover hospital costs.

Vineyards were common around Vienna, as in other European cities. Roughly half of all tax-paying burghers had vineyards for their own use, and the city produced enough to export wine up the Danube as far as Bavaria. But there was plenty left for daily use. Tax records show that the Viennese had an average per capita consumption of 150 liters of wine per year—i.e., half a bottle a day. At the end of the sixteenth century, wine producers in and around Vienna found the conditions worsening, as a petition by Viennese burghers from 1593 makes clear:

> We have learned, what seven years of spoiled wine harvests can do to daily maintenance [i.e., daily survival]. Those who had been seen as especially wealthy and well to do in the cities and markets have grown as poor as all others, and are as impotent as they are. It would have been better for them to stay away from growing wine than to pursue it with such loss to their own provisions.[20]

A succession of bad harvests between 1587 and 1594 had brought the winemakers to their knees. Frequently, no more than half the normal harvest was brought in, and even what could be harvested was so sour as to be almost undrinkable.

The town of Krems, about fifty miles (some eighty kilometers) west of Vienna along the Danube, had been important in the regional wine trade since the Middle Ages, as its splendid Italianate townhouses attest. Around the turn of the seventeenth century, however, two other businesses gained in importance: lead white pigments for paint, and mustard. The production of both required vinegar, which was available in increasing quantities, as the must (the crushed grapes) was not of adequate quality for wine. At the same time, a whole new system of cellars was excavated beneath the town. What may seem paradoxical—expanding the cellars when harvests were cut in half—is revealed as a sound business decision. Much of the wine was so sour that it had to be aged for years before it lost enough of its acidity to be enjoyed at all.

In a warm year such as 1586, some forty-five hectoliters (almost twelve hundred gallons) of wine per hectare could be harvested (a very large amount by today's standards, and not indicative of high wine quality). In the following year, the average harvest had collapsed to seven hectoliters (about 185 gallons) per hectare: "Anno 1586 was a great vintage, after that, a great, cold and strong winter came and the vines were all killed by frost,"[21] as one contemporary, Georg Payr, noted.

West of the German Habsburg lands, in the Alps, plants were also affected by the cold, as was the entire landscape. Alpine glaciers as far as Switzerland and France expanded, devouring summer meadows, farmsteads, entire villages, and, to the horror of local churchmen, even small chapels. But the cold also may have been partly responsible for a phenomenon that has baffled scientists for four centuries.

Alpine forests were known to produce excellent wood for musical instruments, and luthiers from the Bavarian town of Füssen, on the north side of the Alps, to Cremona in the south, benefited from their proximity to the best kinds of pine. Masters such as Antonio Stradivari and members of the Guarneri and Amati families used this wood in building their instruments, which were of particularly fine quality between roughly 1650 and 1750.

Today these instruments are more highly prized than ever, and while the astronomical sums paid for them are partly due to speculation or simple rarity in the market, there is no denying that some of these instruments have an exceptionally fine, rich, and varied sound. What is the secret of their quality? Countless scientists, musical amateurs, and even downright cranks have attempted to solve this mystery. Is the sound simply in the great age of the instruments, has it evolved over generations of use, does it lie in exceptional craftsmanship, in the form or the varnish, or some other secret the old makers took with them to their graves?

In 2003, Henri Grissino-Mayer of the University of Tennessee and Lloyd Burckle of Columbia University published a study in which they suggested that the factor most likely responsible for the exceptional characteristics of golden-age Italian stringed instruments may be that, during the Little Ice Age, the trees were taking longer to mature and grow, and that the wood, with closer tree rings and denser structure, has better sound qualities and more intense resonance. Since the wood used for building instruments of high quality was frequently dried and aged for up to a generation, the trees used around 1700 would have been felled no later than 1680, and thus would have grown and matured during the peak of the Little Ice Age.

The intriguing thesis put forward by Grissino-Mayer and Burckle has been criticized by fellow scientists, who have pointed out that there are also outstanding instruments from this period not built from wood with particularly close tree rings, and it is true that this finding leaves open a number of questions. Why, for instance, did violin makers working at the same time and with similar wood but in different countries not achieve the same high standard? A final answer will have to take into account more factors, but the study reminds us how vital are the material preconditions for all craft, all culture, and all art. The Little Ice Age had its own genre of painting, its own poetry, and its own fears of apocalypse. It may also have had its own instruments, its very own sound.

THE LIGHTS GO OUT

While Italians could rejoice in producing some of the world's finest musical instruments during the Little Ice Age, the variable climatic conditions mainly brought hardship and uncertainty, especially during the cold 1590s, which saw famine and unrest throughout northern Italy. Harvests were low even as far south as Apulia and Sicily, with the inhabitants of northern regions such as Tuscany and the Veneto having to look beyond their own territories for food. The Venetian senate and the Grand Duke of Tuscany sent delegations to Danzig, Amsterdam, and Hamburg to negotiate emergency grain deliveries. During the following years, some two hundred thousand tons of Baltic grain were sold to northern Italy every year.

Often, the bulk of the deliveries included rye, which in Italy was deemed an inferior grain. The population complained about the "foreign bread" yet had no choice but to eat it.

This foreign bread was indicative of changes that extended far beyond food quality or commercial considerations in general. Social historian Christopher Black draws the connection:

> [T]he crises of the 1590s . . . had a wider impact. That decade saw as elsewhere in Europe, years of poor harvests under adverse weather conditions, famines, recessions in textile industries. The hunt for new food supplies brought in Dutch and English shippers, transporting grains from the Baltic. Their entry into the Mediterranean markets opened up opportunities and entice-ments that were not forgotten; so they remained to challenge the economic trading activities of the Venetians in particular, and to add to the problems of piracy on the high seas.[22]

This hastened the demise of Italian trading dominance in the Med-iterranean. With the gradual mapping of the world in the fifteenth and sixteenth centuries, northern Europeans had become more interested in overseas trade than in competing against the powerful Italian states in the Mediterranean. Now, with Italy weakened and dependent on trade with North Sea ports, the ascendancy of the northern merchants and their interests became unstoppable, in the Mediterranean as elsewhere.

Meanwhile, harvests were not recovering. Many regions had neither the financial means, nor the geographical or political access to rivers or ports, to warrant large-scale imports of grains. After the bad harvests of the 1590s, storehouses in many cities stood empty and grain prices climbed dramatically. The average price of bread became higher than at any time in living memory. Discounting other inflationary pressures, prices for wheat rose continually between 1550 and 1590, only to surge during the fol-lowing decade. In the markets of Cologne, wheat cost two and a half times more; in the great Les Halles food markets of Paris, as well as in Tuscany and Spain, it was three or four times as much; and in Vienna, even five times.

In good years, the prices for grain and other basic foods would decline, but the constant seesawing of the cost of living, combined with occasional famine and hefty overall inflation, created a pervasive sensation of uncertainty. Whereas before the Little Ice Age famines and epidemics had afflicted Europeans roughly every ten years, now these drastic events were occurring almost three times as often.

⋈

IT COMES AS little surprise that many people turned to the Church in search of answers to this profoundly disturbing crisis, caused, as all observers agreed, by remarkable and baffling changes in the way nature itself was behaving. It was all the more unsettling because it seemed that God's own Creation was refusing to nourish his premier and most beloved creature, man. The Little Ice Age was gradually rendering the familiar unfamiliar, withering not only local agriculture, the very base of Europe's economic system, but also much of its social cohesion. Cracks in the economic system affected the entire social and political edifice, until the fissures in its elaborate, still largely medieval façade—created over centuries by men of religion and artists and philosophers—could no longer be hidden.

To remain standing, this edifice would eventually have to be rebuilt on new, more solid foundations. At the turn of the seventeenth century, however, Europeans found comfort in looking for explanations in the truth as they had learned to see it: in religion and in magic. The interpretation of natural events, whether sudden frosts, comets, or floods, was entrusted to theologians, and their answer was clear: God had sent these perils to punish his wayward creatures or to try their virtue; they would need to repent of their sinful ways. The response to natural catastrophes still seemed to be a matter of morality.

And yet, not all Europeans were satisfied with the old religious answers. Substitutes proliferated, much to the fury of the Catholic

Church's Inquisition. This was a time of messiahs, seers, occultists, witches, magicians, and charlatans, of alchemists and soothsayers, all offering alternative reasons for the evidently disturbed workings of the universe. Monarchs such as England's Elizabeth I and Habsburg Emperor Rudolf II even employed court magicians, allowing themselves to be guided by their horoscopes and pronouncements. Books and pamphlets on magic and necromancy did a brisk trade.

With the natural order apparently shaken, many pious or popular commentators also predicted an imminent Last Judgment. Every remote rumbling of thunder was interpreted as the first harbinger of apocalypse—the distant, clattering hooves of the Four Horsemen bringing death and destruction in their wake. From this point of view, it made sense to conclude that if nature was simply following the divine command by withholding its blessings, Christians would have to restore the balance, making amends in their individual and communal lives in order to return things to their previous, harmonious state. Repentance and personal forms of piety (which even in Catholic areas also drew inspiration from Protestant habits) rapidly gained in popularity from Spain to France, southern Italy, and the Alpine region. Dramatic large processions were organized in a show of appeasement, to implore the Lord's mercy. Holy relics were carried aloft, with flagellants following, whipping themselves repentantly until the blood streamed down their backs.

In his great *Don Quixote*, first published in 1605, Miguel de Cervantes describes just one such procession, whose participants are accosted by the Knight of the Woeful Countenance (Don Quixote himself), who has mistaken the Virgin Mary statue being carried by the faithful for a noble damsel in distress, in need of rescue by a strong knightly arm.

> Don Quixote rising to his feet and turning his eyes to the quarter where the sound had been heard, suddenly saw coming down the slope of a hill several men clad in white like penitents.

The fact was that the clouds had that year withheld their moisture from the earth, and in all the villages of the district they were organizing processions, rogations, and penances, imploring God to open the hands of his mercy and send the rain; and to this end the people of a village that was hard by were going in procession to a holy hermitage there was on one side of that valley.[23]

Of course the noble Don Quixote tries to rescue his damsel, and of course the story ends badly for him. Cervantes, however, used a very common event for this story: Processions praying for rain were frequent in Spain, as drought made normal life all but impossible for local farmers. Even in Barcelona, at the coast, two or three rain processions were held annually around 1600. In a religious and superstitious age, divine intercession seemed the best, indeed the only hope for restoring nature to the harmony it had so clearly lost.

WITCHES AND SPOILED HARVESTS

In the north of Europe, and especially in German-speaking areas, the accumulation of bad harvests and the constant fear of famine and illness caused an increase in a particularly cruel collective hysteria: witch trials.

Toward the end of the sixteenth century, Nördlingen, in southern Germany, was a wealthy market town. Most of its inhabitants had thrown their fate in with the Reformation, and Nördlingen had become Protestant. Centuries of trade and a strategically advantageous position meant that the leading burghers not only could afford fine houses for themselves, they also had been able to finance a grand town hall. This was the workplace of Peter Lemp, the town's paymaster and one of its most respected citizens. He owned a house on the Wine Market Square, where he lived with his wife Rebecca and their four children.

When the "crazy woman" Ursula Haider claimed openly to have had sex with the devil and to have murdered a child, she was arrested and handed over to special investigators. After an initial interrogation, she was accused of witchcraft, a serious, indeed a capital offense. Under torture, Ursula Haider began to accuse other women in the town of similar offenses. One of those accused was Rebecca Lemp. On July 1, 1590, while her influential husband was away attending to his business, she was arrested and accused of being a witch.

Initially, this prominent prisoner was treated with some circumspection, but when she refused to admit to witchcraft, the judges ordered her to be "put to the question"—that is, tortured. As regulated by the procedures of the day, which forbade the spilling of prisoners' blood, she was tortured "bloodlessly," first with thumb screws and "Spanish boots"—wooden planks attached to the lower legs and made ever tighter by driving wedges between them until the shinbone was crushed. When she still refused to admit any wrongdoing, she was stretched on the rack, then suspended from the ceiling by her hands, which were tied behind her back.

Returning from his long absence, Rebecca's horrified husband did everything in his power to have his wife freed, but without success. Twice she managed to have notes smuggled out to him. The first of these, written after almost six weeks of imprisonment and torture, reveals her fear, but also her as-yet-unbroken hope:

My heart, my darling, be without worry. And if 1000 of them accused me, I am innocent, or may all the devils come and tear me apart. And if they want to question me in terrible ways, I have nothing to confess, even if they tear me in 1000 pieces. Be without worry, by my soul I am innocent. When I am tortured I fail to believe it, because I am just. Father, if I am guilty of something, let me not see God's countenance in all eternity. If they don't believe me, God the Highest will see to it and send a sign. For if I am made to endure this terror long there is no God in the Heavens. Give testimony on my behalf. You know my innocence. For God's sake, do not leave me alone in this terrible place.[24]

Shortly after Rebecca Lemp wrote these lines, she finally collapsed under renewed torture and told the judges everything they wanted to hear. She had been the devil's lover, she claimed. The next and last letter she managed to smuggle to her husband reveals a woman who had lost all hope of either God's kindness or her own earthly vindication. Knowing what to expect, she now asked her husband Peter to send her poison:

> O my precious darling, I am forcibly taken away from you in spite of my innocence, God may receive my laments forever. I am accused, I am made to talk, I am tortured. I am as innocent as God in Heaven. If I knew a single tiny point of any such thing, I would that God closes his Heavens to me. O you, my beloved darling, how heavy is my heart. Alas and alack for my poor orphans. Father, send me something that I may die, otherwise I will go mad from the torment. If not today, do it tomorrow. Write to me now.
>
> Wear the ring for my sake. The necklace make into six parts. Let them be worn by our children under their shirts, for all their lives. O darling . . . I am taken from you by force. How can God tolerate it? If I am bad, let God show me no mercy. But now I am overcome by injustice over injustice. Why does God not want to hear me? Send me something, otherwise I have to speak false. I will forfeit my soul by doing so.[25]

The poison failed to reach her in time. On September 9, two months after her arrest, Rebecca Lemp and four other condemned women were taken to a hill in front of the town and burned alive at the stake. After the executions, the town council paid for a banquet to celebrate this victory over the forces of darkness.

The case of Rebecca Lemp was one of the first in a wave of witch trials held from the late 1580s onward, mainly in southern and central Germany, Switzerland, and eastern France. The protocol of her interrogation shows that under torture Rebecca had admitted not only to sleeping with the devil but also to inflicting harm on people and animals with black magic.

Lemp's case was in some ways not typical, because she was a wealthy and respected forty-one-year-old literate woman. Most of the accused tended to be older, poorer women on the margins of society, often widowed and without children to defend them, as well as some men. The accusations leveled against Rebecca Lemp, however, are entirely characteristic of the witch trials of the time: practicing immoral acts with the devil and harming animals, pregnant women, children, and cattle. "Most importantly, however," writes historian Richard van Dülmen, the presumptive witches and warlocks were accused of having summoned "great plagues of mice which ate the corn, as well as bad weather with hail. Even a devastating snow fall in summer was thought to be their doing."[26]

In 1588, the year the Spanish Armada was destroyed by an Arctic storm, icy winds had raced all over Europe. The scholar Victorinus Schönfeld noted that the year of 1589 in southern Germany had been "cold, dark, hard and windy," with long periods of frost and rain. In other regions, almost no rain had fallen for weeks. In the Prussian town of Stendal, Daniel Schaller documented even earthquakes, and a chronicler in nearby Aschersleben wrote the following about his own town:

> 1589 was oppressively and burningly hot in June and July. The workmen, who almost died because of their great thirst and were not careful when they drank [i.e., drank too much, too suddenly] fell ill. . . . The constant heat caused the water mills to stand still. 1590 on 16 September people noted here and at other places in Germany a violent shaking of the earth, which lasted for 14 days. Doors and windows shook, the tower swayed . . . and even the hams strung up high under the ceiling in the apartment of the town piper began to move according to police. Again, as in the year before, there was a long period of drought, as well as a lack of water and flour. As the water mills were not working, people had to harness horses to millstones. As a result, people were falling ill and prices kept rising. The pine forest

caught fire again and only after great exertions could the flames be stopped after four weeks of continuous burning.[27]

Perhaps it is not surprising that a Prussian author writing under the pseudonym of Adalbertus Temopedius felt confident in predicting that the end of the world would occur during the week before Easter of 1599. Pastor Daniel Schaller argued against this daring prediction, but not against the opinion that "the end of the world and the Last Judgment / are no longer far and distant / but are close by, in front of the door."[28]

The German town of Nördlingen was also struck by this toxic combination of severe weather and fear. Food prices had doubled, and it was a rare occurrence to see the sun even during the summer months. The suspicion that all this was due to witchcraft was widespread. Even so, the authorities had initially been reluctant to conduct witch trials, and had resisted popular pressure to do so.

When entire fields were destroyed by hail near the town of Werdenfels in the summer of 1581, peasants attempted to convince the judge of the nearby town of Garmisch to hunt for the witches

Every catastrophe needs culprits. In central Europe, disastrous harvests were commonly followed by a wave of witch trials.

who had wrought this damage, but the local prince refused to let the judge proceed. Other municipalities had made similar decisions. Witches had already been prosecuted in Luxembourg and France, in northern Italy and Catalonia, but Bavaria, Württemberg, and other German regions initially had been slow to follow suit.

This changed around 1590. In order to understand this shift and to gain an impression of the atmosphere in a town like Nördlingen, it helps to consult Christoph Schorer's *Chronicle of the Town of Memmingen*, which paints a vivid picture of extreme weather and the extreme reactions to it. Schorer collected news, rumors, and reports from various sources and collated them into one panorama. It is remarkable how doom-laden and apocalyptic these items of early (frequently fake) news were. He wrote about witches being burned, portentous astrological phenomena, babies born with terrible deformations, and tragic and mysterious deaths. It was almost as though all Creation had risen up in mutiny:

1589. On 20 May the weather was terrifying / and people thought they would all perish.

1590. In that year many wicked persons were burned at the stake in the surrounding land.

It was a warm summer / and everything ripened well.

1592. On 28 March Hans Kleiber Metzger's grown daughter hanged herself in the house of her father. She was put in a barrel and sent downriver.

On 27 March the night was light / and the heavens open.

We saw many things like that, also on 11 and 12 April that year.

On 3 July the sun was seen bloody red.

On 3 December a woman in Worningen gave birth to a child with two heads and 4 hands. It was brought to the town hall on the fourth of December / and was depicted by master Abraham Werlin the church warden / who was a painter.

1593. A hot and dry summer / and the crickets ate everything in the fields / resulting in a surge in prices.

On 19 July when people were going to church / they saw here
three suns and a rainbow.

20 October people saw the heavens opening.

1594. On 29 July at 5 and 6 in the afternoon there were two terrible
thunder storms / In Steinheim lightning struck the farm of
Juncker Lutz von Freyburg / burning a cow and two calves.[29]

In Schongau, the witch trials had begun earlier, in 1588, after a
catastrophic hailstorm on July 26 had destroyed the harvest. Hysteria had grabbed the entire area. In Nördlingen, with some ten
thousand inhabitants, more than thirty other women and one man
were burned at the stake within four years. In Bamberg, the witch
persecutions had claimed more than three hundred victims by the
middle of the seventeenth century; in Freiburg, fifty-three; in Würz-
burg, some eleven hundred. Fear of witches and black magic had
always been part of popular culture, but now it grew to monstrous
proportions wherever bad harvests made people anxious and at a
loss for alternative explanations. In the tiny state of Bilstein in West-
phalia alone, twenty-one people were burned as witches in 1590,
and in the nearby little farming town of Lemgo, 272 had been executed by 1600.

Historians are still seeking explanations for why the witch persecutions were especially cruel in the years between 1588 and 1600
and then again between 1620 and 1650. Religious tensions certainly
played a role, but the correlation among extreme weather events,
ruined harvests, and waves of witch trials asserts itself most force-
fully. The geographical distribution of the trials, however, cannot
be explained through this correlation alone.

Some 110,000 witch trials occurred at this time in Europe,
roughly half of them ending in conviction and execution. Courts
in Catholic areas acquitted slightly fewer defendants than those in
Protestant areas, but the documentation of the trials is incomplete
and raises many questions. Frequently, the trial files were destroyed;
in some places, no records were kept at all. Still, it is estimated that
more than fifty thousand people were burned at the stake or exe-

cuted in other ways—half of them in Germany, a quarter of them in France, Switzerland, and the Netherlands, and the rest throughout Europe, from Poland to Portugal.

Not all witches, by the way, were regarded as evil and harmful. The Italian historian Carlo Ginzburg has researched the phenomenon of the *benandanti* in the northern Italian Alps around the end of the sixteenth century. *Benandanti* were male guardians of plenty, warlocks, who would metamorphose into animals at night and fly to the fields in order to defend them from wicked female witches intent on destroying the harvest. As the documents show, it was particularly important in the rugged and barren Alpine regions to keep and defend the fruitfulness of nature.

Geoffrey Parker, one of the leading climate historians working today, describes the connection between high-altitude farming and a particular culture of good and bad witchcraft. His thesis also goes some way toward answering the question about the geographical distribution of witch trials. The most intensive persecutions not only followed hard on the heels of poor or spoiled harvests, they also were particularly frequent in marginal areas that were more vulnerable to changing weather conditions. Farmers in regions such as Scandinavia, Scotland, and the Alps, which were in any case not always warm and sunny enough to ensure good yields, arguably had more to lose than farmers in areas with more advantageous climates. In Luxembourg, eastern France, and southern Germany, the unseasonable hail and endless rain affected the production of not only grain but also wine, which was crucial not only for daily use but also for export. In both cases, the economic underpinnings of the local societies were gravely threatened by extreme weather phenomena.

THE TRUTH IN THE STARS

Toward the end of the sixteenth century, the pervasive impression that nature was out of kilter and that *ruina mundi* (the ruin of the world) was imminent led to an intensive intellectual search for new explanations and patterns of interpretation. It was thought that the discovery of nature's cosmic patterns might possibly bring some kind of power over the hidden mechanisms of Creation.

In an era with no formal scientific method to channel the rules of knowledge acquisition, this search took forms that may seem extraordinary to us but that were quite normal by the standards of the day. "Knowledge" still flowed mainly from religious sources. Religious authorities—the pope and the Catholic Church, as well as Anglican, Huguenot, Calvinist, Lutheran, and other dissenting ministers—led a fierce, Europe-wide battle to maintain the Bible as the only source of all knowledge, including knowledge of the natural world. This meant in practice that every natural observation had to be corroborated with biblical quotations and made to conform to the ancient stories of the Creation, the Flood, and more.

Every earthquake, every volcanic eruption, and every storm was interpreted as an expression of divine will, an articulation of the Creator's displeasure, a punishment for human wickedness. Theological interpretations of climatic events were popular and frequently widely disseminated in print. Indeed, weather sermons became a minor literary genre of their own. The theologian Johann Georg Sigwart was one of the proponents of this phenomenon, as we read in a sermon he delivered in Tübingen in 1599:

> The Almighty has exercised his merciful will here in Tübingen both through price rises and hunger and through the plague, which he has visited upon us so strongly during the year, and a great number of young and old people were taken from us: I have held three sermons on this already, one for every affliction . . . to show what punishment it was, and what had been its cause, and by which means they might be avoided, provided the world will stand much longer: so that every man arrives at honest repentance, a Christian life and good deeds in order to move our Heavenly Father to remove them altogether, or at least to make these punishments less severe.[30]

In the face of the otherwise inexplicable, sermons like these, insisting on a direct causal link between bad behavior and bad harvests, were still very popular, but by now the men of God no longer had the unquestioned authority that they had once possessed. The religious Crusades of the High Middle Ages had reintroduced Europeans to the aesthetic and intellectual riches of Greek and Roman antiquity, kept alive over the centuries of Europe's "dark ages" by Arab scholars and astronomers in the Middle East. The discovery of new continents from the fifteenth century onward vastly expanded the Europeans' understanding of the world, as did their widening network of trading contacts, and new astronomical theses that dislodged the Earth from its previously given position at the center of the universe. All these events and discoveries challenged the worldview of the men of God—explicitly when it was safe to do so, and at times not even then (more about some of these brave souls later), and always implicitly, as scientific knowledge steadily grew.

For the elite, the rediscovered learning of the ancients offered a powerful alternative foundation of knowledge. The ancient philosophers were not deemed equal to the Gospels in authority, but they were nonetheless now often quoted at length, and this allowed literate people to think and speak about the natural world without recourse to the Bible. A dialogue with Aristotle and Plato, Herodotus and Pliny made it possible, in theory at least, to escape the heavy millstone of Church teaching.

A central work for this new way of looking at the world, and for communicating with secular thinkers across the centuries, was a materialist poem from the first century CE, *De Rerum Natura*, by the Roman philosopher Lucretius. Very little is known about him, apart from the slander heaped on him by the Christian Saint Jerome, who insisted, three hundred years later, that the poet and thinker had died by his own hand after having been driven mad by a love potion. Considering that Jerome himself espoused the most dour and ascetic form of Christian life and worship, his remarks seem very much like a posthumous hatchet job to discredit one of the loveliest and most powerful poetic works extant in Latin. Still, his tactic appears to have been successful for a considerable period. Lucretius was almost completely forgotten until a moldering manuscript of his only surviving work was rediscovered early in the fifteenth century, as Stephen Greenblatt recounts in his book *The Swerve*.

Lucretius describes a world in which the gods no longer have any role to play in earthly matters, where all human striving is oriented toward minimizing human suffering and maximizing pleasure, without becoming a slave to the passions. Humanists already knew these ideas from such ancient philosophers as the Stoics, the Epicureans, and the Cynics, but the rediscovery of this beautiful and sweeping narrative caused a sensation in European intellectual circles, and soon *De Rerum Natura* was reprinted and read across the Continent.

At a time when nature appeared to be in full revolt and theologians were blaming human sins for natural disasters, Lucretius provided an indispensable alternative view of the world. Everything in nature happens according to immutable laws, he wrote, free from the intervention of the gods. Natural events have no moral significance, the soul dies with the body, and it is meaningless to fear death because there is no Last Judgment and no hell, only nothingness. What remains is the cultivation of one's own being in harmony with the laws of nature and through the power that drives, permeates, and creates everything: eros. The first verses of the great poem are therefore fittingly a hymn to Venus, the goddess of love and sovereign ruler over all beings:

> Mother of Rome, delight of Gods and men,
> Dear Venus that beneath the gliding stars
> Makest to teem the many-voyaged main
> And fruitful lands—for all of living things
> Through thee alone are evermore conceived,
> Through thee are risen to visit the great sun.[31]

In a universe still ruled by theology, a universe in which all sensuality was deemed sinful, this serene vision resonated powerfully, as an increasing number of thinking people were looking for explanations beyond the teachings of the Church.

Lucretius marked one materialist response to nature, but there were many others jostling for power. Through the work of Poggio Bracciolini and others, the discovery of valuable ancient manuscripts had widened the repertoire of what could constitute knowledge. Mystical traditions of other religions—particularly the Jewish Kabbalah and Muslim teachings on astrology in the tradition of Averroës (Ibn Rushd)—not only were popular in esoteric circles but also were devoured with enthusiasm by those with diverse beliefs. More or less fanciful translations and commentaries that reflected very different standards of knowledge about other languages and traditions made the rounds in Europe and were used to posit new explanations and worldviews.

These mystical models of the world created an all-encompassing system of meaning and symbolism whose venerable age made them seemingly unassailable. At the same time, another equally old method of gathering knowledge reappeared in the debate: empirical observation. Scholars, writers, churchmen, teachers, and businessmen began to record their own notes about the natural world, its changes and its apparent laws. Often these observations were intermingled with biblical prophesies, reports of miracles, and mystical predictions, but they tended to be systematic, they collected useful data, and at times the data were even correlated with observations by others. The first shoots of a scientific method began to be visible among the towering trunks of religious tradition.

The Protestant pastor David Fabricius (1564–1617) was one of these searchers for a truth that was less mystical and more down-to-earth. In the tiny northern German village of Resterhafe, he made notes about the weather every day and observed the stars at night, as often as the gray Frisian skies would allow. In 1611, his son, a student at Leiden University in the Netherlands, gave him a telescope, one of the first to be used in the German lands. Even before receiving this precious instrument, however, Fabricius had already made some important astronomical discoveries; he had corresponded extensively with astronomers such as Tycho Brahe in Denmark and Johannes Kepler, then in Graz, and he wrote the first scientific text about sunspots. Even though he was a cleric, he did not use his observations to calculate the date of the apocalypse or to avert divine punishment, though he did calculate horoscopes for himself. Astrology, after all, was still deemed part of astronomy. We will encounter Fabricius again.

DOCTOR FAUSTUS

The empirical, experimental approach to nature being used by early scientists such as Fabricius was still only one of many possible kinds of knowledge. Many if not most scholars regarded empirical deduction as useful, but certainly not the best method of determining how things worked. Magic and the interpretation of biblical texts seemed to be at least equally promising. One of the most famous learned men of his day, John Dee (1527–1608), is an English representative of the Europe-wide profusion of Faustian figures who were attempting to penetrate the mysteries of nature through symbols and incantations.

The son of a mercer, a cloth merchant, Dee showed his intellectual brilliance at a young age. He was sent to Cambridge, but he also studied in Louvain and Brussels before returning to England, where he appears to have oscillated between his father's business and a growing interest in mystical calculations. This latter interest even earned him a trial for treason after he had drawn up the horoscope of Mary Queen of Scots and Henry VIII's daughter, the Princess Elizabeth. When Elizabeth ascended the throne in 1558, Dee became her close adviser. Even the date of the new monarch's coronation was chosen according to his calculations. In his new position of power, Dee discovered an interest in politics, defending England's claim to a colonial empire, a prophetic feat of sorts. It was he who coined the term *the British Empire*.

Dee's writings cover many subjects, but his impenetrably obscure

mystical tract *Monas Hieroglyphica* (1564) immediately made his name
as a magus and expert on the Jewish mystical tradition of the Kabbalah and other branches of hermetic knowledge. Still, Dee was a
worldly magician. At the royal court in London, he had made powerful enemies, so he dedicated his magnum opus to the Habsburg
Emperor Maximilian II, thus creating a new source of patronage
for himself. Meanwhile, his mystical work was also progressing.

Dee's copious diaries show that he had little interest in meteo-
rology or systematic observation. They consist of notes about his
family, endless accounts of income and expenses, including pay-
ments to his servants, reports about visits to other scholars, health
problems ("had some wambling in my stomach"), quasi-scientific
experiments, and his relations with Elizabeth I and, later, Habsburg
Emperor Rudolf II. The few times the weather does intrude into
his personal writings, it is to record extraordinary bouts of cold and
high winds during the later years of his life, particularly the 1590s.
When Dee visited Prague, he and his horse were engulfed by deep
snow. On January 15, 1590, he noted that a "terrible tempest" had
hit England. Storms also made an appearance in the pages devoted
to October 1591 and September of the following year: "tempes-
tuous, windy, clowdy, hayl and rayn, after three of the clok after
none."[32] Even June 1596 was "wyndy and rayny," and on June 25
there was "thunder in the morning, rayne in the night."

A rainy summer could not deter the scholar from his obsessive
book collecting, nor from his investigation of the secrets of Creation. Although Dee was famed as a mathematician and a man of
wide-ranging erudition, his particular interest, and that of his powerful patrons, was a special and secret area of knowledge, the occult
sciences and magic.

By 1580, Dee had lost a great deal of his influence at Elizabeth's
court. He now devoted himself more intensely to the study of hermetic philosophy, which he pursued with a determination bordering on the obsessive. In 1587, Christopher Marlowe published his
tragedy *Doctor Faustus*, inspired not only by the eponymous folk-
tale and its German historical personification, the alchemist Jörg

Faust, but also by the queen's former adviser and court magician, Dr. John Dee.

By this time, the magician himself was seeking to establish communication with "the other side," especially with the angels supposedly in residence there, but initial results were discouraging. This changed with the arrival in his life of a young man named Edward Kelley, probably a former student at Oxford University, with a colorful career already behind him. Witnesses reported that Kelley always wore a cap over his long hair; it seems that this was to hide the fact that one of his ears had been cut off—then a common punishment for forging coins.

Kelley proved to be just the collaborator Dee had been needing. Among the young man's many gifts was supposedly that of being a spiritual medium. The angels, or so he claimed, were speaking to him directly, in their own language. Together, the two men held séance after séance, during which Kelley revealed the orders of angels, archangels, seraphim and cherubim, the cartography of the heavens, and the outlines of secret kingdoms. Hidden powers, it seemed, were communicated to him through a numerical code that only he could decipher, or in the language of Enoch, the occult tongue of the celestial realms that had so far proven difficult to translate into plain English. The contours of the messages, however, were clear: All signs pointed toward a *renovatio mundi*, a divine judgment that would usefully also reveal Kelley and Dee as legitimate prophets of the Lord.

In Elizabethan England, Dee soon found that his spiritual quest was beginning to isolate him from his more pragmatic countrymen. On the Continent, however, seers and magicians were in great demand, so the magician and his medium accepted the invitation of a Bohemian nobleman to visit Prague. Here they were to demonstrate their occult arts before one of Europe's most powerful men, the Habsburg Emperor Rudolf II, whose penchant for alchemy and other forms of occult knowledge was by now legendary. The melancholic emperor was famed for his vast collections of beautiful and strange objects from around the world. The idea of a secret code

The Magus of the North: John Dee sought to formulate
mystical and magical explanations for natural events.

of Creation hidden in the thousands of objects crammed into his
Hradschin Palace, a code that might explain them all, and with that
the meaning of the universe, was irresistible to the emperor—so
much so that he was known to spend days completely alone with his
collections, engaged in obscure experiments, while the business of
state was left in abeyance.

Rudolf was interested in the two Englishmen who claimed to
be able to perform miracles, but he was not as gullible as they or
others may have thought him. A delighted John Dee confided to his
diary that His Majesty had bestowed on him a necklace that was valued at three hundred ducats, but the emperor's patronage remained
sporadic all the same. The two magicians therefore began a restless

life of travel through Central Europe in search of other patrons—always with great hopes and even greater expectations, always looking for money and recognition, and frequently dogged by misfortune or pursued by disappointed aristocrats who had invested time and money in them, receiving nothing but esoteric-looking codes and vague pronouncements in return.

Whatever Edward Kelley's motivations may have been, John Dee was not simply a charlatan. As the years went by, he became ever more obsessive in his search for the hidden truths of the universe, and soon his program of almost daily séances with the young self-declared medium would be stretching for hours. Eventually, Kelley grew bored with it all, or perhaps he simply thought it was time to take what he could while the going was still relatively good. Through his lips, the angels began to order that the two men should share all their possessions, including their wives. Dee was horrified at first, but he resigned himself to the celestial fiat. Soon afterward, he left Bohemia for England, where, nine months after his arrival, his wife gave birth to a healthy boy.

Kelley and Dee would never meet again; the conversations with the angels had come to an end. Dee attempted to locate other mediums, but he found none who appeared to be as gifted as Kelley. Now his life's work seemed in danger. On his return to his country estate, he had found his vast and beloved library vandalized, and many of his valuable books and scientific instruments stolen. Queen Elizabeth, his former patroness, was dead, and he was no longer welcome at court. During the icy winter of 1608, John Dee died at the age of eighty-two—isolated from society, surrounded only by the remnants of his library.

While the old magician's reputation and life had come to a sad end, his former protégé was still playing on the hopes and greed of his aristocratic patrons—and now with considerable success. Kelley's promise of being able to create gold with the help of an alchemistic tincture finally reawakened the interest of the notoriously indebted Emperor Rudolf, who knighted the virtuoso occultist in 1590 in the expectation of eventually gaining alchemically created vast riches. Rudolf gave his court alchemist a more than comfortable living allowance, but as time went by and Sir Edward's experiments failed to produce even a single nugget of gold, the emperor became impatient.

When Kelley was accused of having killed a Bohemian nobleman in a duel, he was imprisoned, then released and reinstated in his position, only to be sent to jail again, this time without hope of regaining his freedom anytime soon. The magician's situation seemed desperate, even though the emperor had been more than generous to him. Watched and often ridiculed by more skeptical courtiers, Kelley was humiliated. Kelley knew he could not hope for clemency, so, in 1598, he decided to flee. His amazing luck, however, had finally run out. Leaping from a high wall, he broke his leg and died soon afterward, probably from infection.

John Dee and Edward Kelley seem like characters from a play, at a time too distant for us to comprehend fully. But while Kelley was probably just a charlatan, Dee undertook his intellectual voyages into the unknown with the greatest seriousness. He was a kind of Don Quixote of intellectual history, lost between different epochs and horizons. In spite of the dramatic reversals of fortune he saw and suffered, his biography is not unusual for a growing number of men at this time: scholars, artists, jurists, and engineers looking for patronage, opportunities, and knowledge, who frequently understood that a little theater and a few judicious promises were keys to gaining the attention of the rich and powerful men on whom they depended. A considerable number of these itinerant intellectuals were crisscrossing Europe, visiting libraries and courts, leaving behind them a trail of disappointed expectations and unpaid debts, perhaps, but also of learned writings and exciting new ideas.

INFINITE WORLDS

The Italian philosopher, theologian, and scientific thinker Giordano Bruno (1548–1600) pursued a particularly long internal and exter-

Possible Worlds: Giordano Bruno was burnt at the stake in
Rome for speculating about the nature of the universe.

nal journey, one that led him from southern Italy to the German-
speaking world, France, the Netherlands, England, and back to
Italy. Having visited almost every place of intellectual significance
in Europe, he was forced to leave or even flee most of them—not
least because he could never forgo ridiculing his rivals, who, though
almost unfailingly his inferiors in matters of the mind, were not
always without the resources of temporal power. Bruno was a difficult man, perhaps arrogant but perhaps simply too loudly aware

of his own gifts, for he was also one of the most daringly original thinkers in European history.

While Bruno's theological positions may have less resonance for us today, his cosmological ideas have returned to the center of debate in modern physics. He spoke of a world without beginning and end, consisting of infinite parallel worlds, many of them inhabited by intelligent beings. Within the infinity of this universe, however, he found little or no place for the idea of a divine Creation or a Last Judgment, or for the question of why an omnipotent God and master of infinite universes should have any interest in how the tiny, sprawling organisms in any one of these worlds are living their lives.

Bruno was undiplomatic and uncompromising in discussing his ideas, but after sojourns in Paris, Oxford, Frankfurt, and Zurich, he longed for his Italian homeland. Hoping to be appointed to a chair in mathematics in Padua, he finally returned in 1592. Almost immediately, his pugnacious intellectual attitude led to his being arrested. He was transferred to Rome, accused of being a heretic, imprisoned for seven years, condemned to death for his opinions, and, on February 17, 1600, on the beautiful Campo de' Fiori, Giordano Bruno was burned alive.

THE TOWER OF BOOKS

By the end of the sixteenth century, there was no clear intellectual path among theological, occult, and empirical truths, or methods. As the fates of Giordano Bruno and Edward Kelley demonstrate, it was also a path full of potentially fatal pitfalls. If it was possible at all to come to some sort of conclusion about the nature of the world without falling afoul of the Inquisition, then it could only be far away from the seats of temporal and ecclesiastical power, in the countryside, or perhaps in a tower with a good library, providing not only a refuge but also an opportunity for sustained exchanges with friends long gone.

Michel de Montaigne (1533–1592) was a well-traveled lawyer,

landowner, and former mayor of Bordeaux who had also worked as adviser to the king of France. All of these functions brought him respect, but he was to achieve literary immortality through his devotion to the quiet project of his later life, undertaken not at court or in the bustling city but in his quiet tower in the countryside.

A passionate reader and indefatigable observer of things great and small, Montaigne surrounded himself with friends from various centuries, their voices captured between the covers of their books. In the tower on his estate, he would read and explore his own personality, his perceptions, his follies, his weaknesses and strengths. The result would be published as his *Essays*, and they have long been regarded as the most far-reaching and penetrating study of humanity penned during this era of transformation, and very possibly of any time.

Montaigne was born into an old but minor aristocratic family. His paternal ancestors, the Eyquem family, were wealthy traders who had profited from the economic upturn of Bordeaux, which was then known not so much for its fine wines as for its successful trade in cane sugar and slaves. His mother's family came from Toulouse, and her birth name was Louppes de Villeneuve, a linguistic adaptation of Lopez de Villanueva, which may indicate that the family had been part of the Jewish exodus from Spain and Portugal during the religious persecutions of the fifteenth century.

As a landowner and winegrower, Montaigne was well aware of the changes that were occurring in nature, which were clearly affecting the vineyards of southwestern France. In his essay "Of the Education of Children," he mentions not only the harsh frosts of the time but also the then-current religious interpretation of these events, seen through the lens of the Roman author Lucretius:

> When the vines of my village are nipped with the frost, my parish priest presently concludes, that the indignation of God has gone out against all the human race. . . . Who is it, that seeing the havoc of these civil wars of ours, does not cry out, that the machine of the world is near dissolution, and that the day of judgment is at

hand; without considering, that many worse things have been seen, and that, in the meantime, people are very merry in a thousand other parts of the earth for all this? For my part, considering the licence and impunity that always attend such commotions, I wonder they are so moderate, and that there is no more mischief done. To him who feels the hailstones patter about his ears, the whole hemisphere appears to be in storm and tempest....[33]

Even and perhaps especially in his tower, Montaigne was aware that he was only a tiny part of a large and bewilderingly complex world that essentially remained opaque, even if humans could not resist spinning their stories about it, seeing significance and meaning even where there was none. He understood himself to be part of a culture with particular usages and truths, a tiny facet of a global human landscape of cultures. Unlike many of his contemporaries, he was also convinced that the narratives and truths of his society were simply metaphors for some more basic human experience. He posited that meaning is born when the human desire for structure is projected over the chaos of life lived. This human habit of attributing very human needs to a divine will fascinated him.

Montaigne was pursuing stories from the whole breadth of human grandeur and misery, stories of majesty and sometimes of grotesque stupidity. He was an indefatigable observer of the theater of humanity, gleaning his stories through reading or through direct observations during his travels, when he could leave behind his role as Sieur d'Eyquem for a while and observe the differences between customs and attitudes. As a landowner, he was preoccupied with frozen wines, spoiled harvests, and difficult tenants. As a traveler and a reader, he could allow his mind to roam freely and indulge his hunger for insight, as he himself commented:

This greedy humour of new and unknown things helps to nourish in me the desire of travel; but a great many more circumstances contribute to it; I am very willing to quit the government

Michel de Montaigne explored inner worlds and was
attentive to rational explanations of natural phenomena.

of my house. There is, I confess, a kind of convenience in commanding, though it were but in a barn, and in being obeyed by one's people; but 'tis too uniform and languid a pleasure, and is, moreover, of necessity mixed with a thousand vexatious thoughts: one, the poverty and the oppression of your tenants: another, quarrels amongst neighbours: another, the trespasses they make upon you, afflict you;

> *"Or hail-smitten vines and the deceptive farm; now trees damaged by the rains, or years of dearth, now summer's heat burning up the petals, now destructive winters."*

and that God scarce in six months sends a season wherein your bailiff can do his business as he should; but that if it serves the vines, it spoils the meadows:

*"Either the scorching sun burns up your fields, or sudden rains
or frosts destroy your harvests, or a violent wind carries away all
before it."*[34]

These reflections on meteorological extremes, harsh frosts versus
periods of drought, are taken from *De Rerum Natura,* by Lucretius.
Montaigne, however, also found ample opportunity to add observations of his own. Thanks to his collection of travel accounts and the
curiosity that made him speak to every stranger passing through, he
had a fair idea of the climatic extremes elsewhere:

> . . . and since we are now talking of cold, and Frenchmen used
> to wear variety of colours (not I myself, for I seldom wear other
> than black or white, in imitation of my father), let us add another
> story out of Captain Martin du Bellay, who affirms, that in the
> march to Luxembourg, he saw so great frost, that the muni-
> tion wine was cut with hatchets and wedges, and delivered out
> to the soldiers by weight, and that they carried it away in bas-
> kets: and Ovid,
>> *"The wine when out of the cask retains the form of the cask; and
>> is given out not in cups, but in bits."*[35]

Hard winters were becoming an ever more important factor in the
life of rural populations, but Montaigne noticed that people were
using a variety of techniques to counter the effects of this: "Alexander saw a nation, where they bury their fruit-trees in winter, to
protect them from being destroyed by the frost, and we also may see
the same."[36]

Even in southern France, the vines were freezing, and there was
concern that soon it might be necessary for winegrowers to take
precautions similar to those witnessed by Alexander the Great in
Armenia almost two millennia earlier. Montaigne, of course, knew
he could not be certain about this—or indeed about anything else.
Certainty was not part of his repertoire, or his ambition, and it is
this lack of certainty that still makes his *Essays* essential reading.

He was always ready to appreciate the limitedness of his own horizon of understanding and to respect that others might be in a better position to see and understand aspects of reality that remained dark to him.

For most of Montaigne's contemporaries, this pragmatic attitude, based on modesty and curiosity, remained anathema. To them, the light of Christian revelation shone above everything as the only truth in the world. For him, however, the Bible did not contain any illuminating answers; while he quoted other authors on practically every page of his *Essays*, he did not include a single quotation from the Old or New Testament, from the Fathers of the Church, or from living theologians. His partners in conversation were the great thinkers of classical antiquity, who also allowed him to think freely without having to take a position on theological controversies. His impassioned but also impartial curiosity was directed to all things human, to the many different ways in which people understood and narrated their desires and their fears.

The improvisational, free form of the essay, which Montaigne invented for himself, was the ideal vehicle for observations not only of the world around him but also of himself and his own thoughts, without pressing them into a fixed system or any grand thesis. For Montaigne, good questions were more important than good answers, and he was mainly interested in describing and understanding the fabric of his own life. "It is many years ago," he wrote, "that I set myself as the only goal of my thinking and that I am not looking at anything or investigating anything but myself."

Essays was published in 1580, and several editions came out during the remainder of that century. There were enough wealthy customers—books were expensive—who valued his unblinkered, skeptical humanism. The first English edition was published in 1603. William Shakespeare and Francis Bacon were among its readers.

Montaigne opened the door to a new way of being human, far from doctrinal disputes and revealed certainty. Perhaps the highest achievement of a life well lived, he wrote, is living not for eternal bliss, but in the moment:

When I dance, I dance; when I sleep, I sleep. Nay, when I walk alone in a beautiful orchard, if my thoughts are some part of the time taken up with external occurrences, I some part of the time call them back again to my walk, to the orchard, to the sweetness of that solitude, and to myself.

Nature has mother-like observed this, that the actions she has enjoined us for our necessity should be also pleasurable to us; and she invites us to them, not only by reason, but also by appetite, and 'tis injustice to infringe her laws. . . . We are great fools. "He has passed his life in idleness," say we: "I have done nothing to-day." What? have you not lived? that is not only the fundamental, but the most illustrious of your occupations.[37]

Montaigne found a way of engaging with the world while simultaneously retreating from an increasingly precarious reality. His dialogue with thinkers of the past allowed him to sidestep the sectarian arguments of his day. Only one contemporary was exempted from this rule, a man to whom the writer in his tower dedicated an entire chapter: Étienne de la Boëtie, his intimate friend, who had died of the plague in 1563, at the age of only thirty-three. The two men may or may not have had a sexual relationship, but even if the Eros of their friendship may have remained platonic, he had found in the younger man a perfect complement, a person whom he could love "*parce-que c'était lui, parce-que c'était moi*"—because he was himself and I was myself.

La Boëtie left a single work, which has lost nothing of its urgency in the intervening four and a half centuries. *La Servitude Volontaire* (*Voluntary Servitude*) investigates the nature of power and authority. Power is bestowed, never taken, the young philosopher argued, and it is given by human, not divine hands. No one could exercise power if others would not obey him. The acquiescence of others, their obedience and their faith constitute power itself. When faith in the power of the powerful dies, when nobody listens to the king's command anymore, his power is broken, extinguished like the flame of a fire deprived of nourishment.

The thoughts of the two friends met in one central point: The realization that political structures and even the framework for understanding the world are not determined by some supernatural, greater truth, but rather projected into the world by human beings. What we regard as truth is a product of our own minds. Montaigne knew what he was writing about. His country was riven by religious wars, and both sides—Catholics and Huguenots—claimed to be guardians of God's unique revelation and thus entitled to use violence to further the victory of His truth. As a negotiator between the two sides, Montaigne had firsthand knowledge of the blind hatred engendered by this division.

Now, in his tower, Montaigne wrote about the "new southern France"—Brazil, where French explorers had established contact with indigenous peoples, some of whom practiced cannibalism. Many people took this as proof that the heathens were condemned to suffer eternal pain in the flames of hell, and that only a Christian civilization possessed the authentic keys to truth and to God's mercy. Montaigne argued exactly the reverse:

> Now . . . I find that there is nothing barbarous and savage in this nation, by anything that I can gather, excepting, that every one gives the title of barbarism to everything that is not in use in his own country. As, indeed, we have no other level of truth and reason than the example and idea of the opinions and customs of the place wherein we live: there is always the perfect religion, there the perfect government, there the most exact and accomplished usage of all things. They are savages at the same rate that we say fruits are wild, which nature produces of herself and by her own ordinary progress; whereas, in truth, we ought rather to call those wild whose natures we have changed by our artifice and diverted from the common order.[38]

Montaigne's tower was a place of contemplation but not of isolation. The author's curiosity was never slaked, and he continued to test his ideas in real life. In Rouen, he met three indigenous South

Americans who had arrived in Europe on a French ship, and even if the translator proved to be poor, he still managed to have a brief discussion with these emissaries from a very different world. The Indians, who had also been introduced to the then-twelve-year-old King Charles IX, confessed their surprise at seeing that so many "bearded, well-armed men" were obeying a child, and also that these men appeared to be "stuffed with all kinds of treasure," while others were begging at the door, scrawny and half-starved, curiously accepting of their fate, instead of grabbing the rich men by the throat and burning their houses.

Publication of Montaigne's *Essays* marked the appearance of an attitude that would take centuries to establish itself in Europe and that indeed may never establish itself entirely. In his refuge, Montaigne could write without fear and work at the only project he considered worthwhile: a deep and, as far as possible, unprejudiced investigation of himself and the world in which he lived.

Beyond the confines of his peaceful library, however, disorder ruled. Failed harvests and famines, icy winters and rainy summers continued to make countless people believe that the natural order itself was out of kilter. Bread riots, religious wars, and witch trials created a pervasive feeling of threat and crisis. The old way of the world and the old answers no longer seemed appropriate, and neither seemed to be adequate reactions to the new disorder. More and more contemporaries remarked that burning more witches had failed to bring back bountiful harvests, processions had not stopped floods, prayers had brought no rain, and holy relics had not succeeded in arresting the progress of glaciers. Educated people wanting to understand their era more frequently reached for a copy of Montaigne or Lucretius instead of Luther or the Old Testament. Something was changing. New ways had to be found in order to meet the challenges of a world that appeared to have become unnatural.

THE AGE OF IRON

The forests are stripped of their leaves, the earth lies frozen, the rivers are frozen with the cold. And fog and rain, together with the excess of endless nights, have robbed the earth of its joy.

—MATTHÄUS MERIAN, 1622

HORTUS BOTANICUS

It must have been a remarkable sight. An old, bearded man hardly able to walk, surrounded by flowerbeds in which the rarest plants were growing in painstakingly neat rows. This is not just any garden. It is the *Hortus botanicus* of Leiden University in the Netherlands, one of the world's most important collections of living plants, founded by a famous scholar at the very end of his career. Born in Flanders as Charles de l'Écluse, he was known in learned circles as Carolus Clusius (1525–1609), the greatest botanist of his day. It is he who is inspecting his precious plants, many of which have been sent to him from around the world by scholars, gardeners, merchants, and explorers.

None of his contemporaries had a comparable knowledge about botany. He was born not far from Leiden, his final residence: in Arras in Flanders, where he had grown up as the son of a councillor. The son had been destined for the Church, but amid the religious conflicts centering on the Low Countries, his life took a different direction. Clusius first studied theology and philosophy at the Catholic University at Louvain, then he converted to Protestantism and continued his studies under Philip Melanchthon in Wittenberg, the epicenter of the Reformation. The great humanist scholar had advised his student to follow his interest in nature and particularly in plants, so Clusius took his leave from yet another place of study and headed for Montpellier in southern France, one of the foremost centers of medical learning and what was then called *natural history* or *natural philosophy*.

Traveling remained an important part of the young scholar's life. During the following years, he first visited Spain, Portugal, and Great Britain. Then he became court physician and director of the imperial gardens in Vienna; while there, he was among the first to ascend Austria's Ötscher glacier. He studied the flora and fauna of the Alps, and collected dozens of kinds of mushrooms previously unknown to science on the Hungarian steppes, before settling in Frankfurt and eventually moving to the Netherlands. Wherever he went, he looked for interesting colleagues, rare books, and plants, which he did not dry and press—as was the custom of the time—but transported as saplings, to be planted and grown in his own gardens.

When Clusius became a professor in Leiden in 1593, he was already sixty-seven. Injuries sustained during years of adventurous expeditions gave him increasing health problems, but he ignored his pain and continued working. He designed and built a new garden, corresponded in seven languages with more than three hundred scholars throughout Europe, swapped plants with them, and described rare specimens brought to him by seamen returning from the Americas, Africa, and Asia. He also supervised the work of illustrators, corrected page proofs, propagated and experimented with his plants, and researched.

In addition to all this daily practical activity, the aged scientist also wrote works of stupendous scholarship, works with such ornate titles as *Rariorum plantarum historia: Fungorum in Pannoniis observatorum brevis historia* (1601), and *Exoticorum libri decem: Quibus*

animalium, plantarum, aromatum, aliorumque peregrinorum fructuum histo-
riae describuntur (1605).

As a practitioner witnessing an immense number of new discoveries, spurring the need for research, understanding, and classification, Clusius was also one of the first Europeans to appreciate that a scholar working alone could no longer grasp all knowledge about the natural world, and he therefore had to correspond with many others. "He who wants to know all plants is working in vain," he wrote. "This science must be lifted up by the work and attention of many scholars. You contribute something, I something else and others do the same and thus our descendants will command a more perfect knowledge of plants."[1]

Many of the plants that Clusius discovered or described for the first time and sent out to friends and colleagues across the Continent have altered the landscapes, the gardens, and the diets of Europe and the Europeans, and eventually of the world. He imported horse chestnuts to Vienna from Constantinople, enriched gardens with irises, hyacinths, anemones, and narcissi. His particular interest was a small bulb that had been known in European botanical collections for a few decades, but it was Clusius who introduced it to Leiden.

Legend has it that already in 1562, a small bag of these bulbs had been sent by merchant ship from Constantinople to Antwerp, along with a cargo of cloth. The Ottoman cloth merchant had sent them as a gift to his colleague across the sea. The recipient thought them a kind of onion and gave them to his chef for preparation, but apparently he was disappointed with the result and ordered his servants to throw the rest on the rubbish heap.

When spring came, the rubbish heap was bursting with spectacular, colorful blossoms in the shape of a turban. The merchant rescued them and sent them to Clusius, who had seen these flowers in Spain, but now he could plant them himself. Enthusiastically, he wrote to Ghislain de Busbecq, the Habsburg ambassador to the Sublime Porte (Ottoman Empire), asking him for more bulbs of this magnificent flower. Clusius had cultivated them in Vienna before Emperor Rudolf II had decided to have the imperial gardens razed,

A rare madness: a single tulip could cost
as much as a country estate.

and he had taken them with him when he left. In 1593, when he arrived in Leiden, these flowers had become his main interest. Using a word derived from a Turkish term for "turban," he named it *Tulipa*.

The winter of 1593 was fiercely cold. A correspondent from Norway wrote a despairing letter to his friend in the Netherlands: "Almost all seeds have died, to my great chagrin. *Petum Mas* [the tobacco plant], which an Amsterdam apothecary had sent me last autumn, has succumbed to the attack of winter."[2] Leiden was also in the icy grip of winter, but the tulip bulbs survived the extreme conditions and developed into a small botanical sensation. Clusius sent them as gifts to his friends, but the general public and interested laymen had to pay high prices for the Turkish blooms. His interest was not commercialization but scholarship. The many different colors and their heritage fascinated him, and he wanted to keep the plants in his botanical gardens. Savvy Dutch entrepreneurs, however, had other ideas. They stole some of the bulbs, and soon robust and riotous tulips graced many a Dutch drawing room.

Another plant close to the botanist's heart had also been known to collectors for its beautiful blossoms. *Papas Peruanorum* had been brought to Europe from South America by Spanish conquistadores during the sixteenth century. The English explorer and sometime pirate Sir Francis Drake had also imported some of these bulbs, which were obviously intriguing, even if nobody quite knew what to do with them. They remained a botanical rarity. Clusius had been presented with two specimens in 1588 and had described them in his 1601 work *Rariorum plantarum historia*. He was one of the most important propagators of the bulbs and recommended them not for their beautiful blossoms but for their nutritional value and their sturdiness, capable of weathering even the most inclement conditions. Even when wheat would not ripen, these bulbs did—a great advantage in uncertain times. Still, rural folk were skeptical about the new crop now known as potatoes, and it would be decades before this plant's unique career would take off.

Carolus Clusius died in Leiden in 1609, at the ripe old age of eighty-three. His research and his correspondence had played a crucial role in ushering in a new chapter of scientific inquiry, and the plants he had found were to change the landscapes and even the nutrition of an entire continent. He had corresponded not only with important scholars and mighty noblemen but also with apothecaries and amateurs. Among the people he chose for these intellectual discourses were several women who were part of an international community dedicated to observation and knowledge acquisition. Even his curiosity had a new quality in his day: He studied plants not for their medical properties or their allegorical significance in the grand scheme of Creation, but simply in order to understand their physiology and their places among the other plants. He was not interested in mythology and legends, did not quote authors from classical antiquity, did not use the Bible to claim authority for his opinions. Instead, he submitted them to the scrutiny of colleagues. His interest was to collect, to systematize, to publish. He was one of the first scientists in the modern sense of the term.

REVOLUTIONARY PLACES

Charles de l'Écluse left behind a botanical garden that was to be part of Leiden's claim to international fame and scientific standing, the beacon of a new and systematic determination to investigate nature for its own sake. When the botanist arrived there, the university had been founded less than two decades earlier as a place of learning modeled after Montpellier, where Clusius had discovered and developed his passion for plants as a young man.

The universities in Leiden and Montpellier differed from the majority of European universities in several important ways. While places of higher learning from Naples and Toledo to the Sorbonne, Oxford, and Kraków almost exclusively offered studies in law and theology, together with ancient languages and perhaps a little history of classical antiquity, these two universities concentrated on the emerging natural sciences as well as, in the case of Leiden, Oriental languages, geography, ethnography, and modern history. Even more important, however, neither of these institutions required its students to take an oath on the Bible, pledging allegiance to a particular church. You had to be a Catholic in Bologna or Paris, an Anglican in Cambridge, a Calvinist in Geneva. In Leiden and Montpellier, none of that mattered—at least not officially.

Leiden was open to all. Soon after the 1575 foundation of its university, the town, which had grown wealthy by producing broadcloth from Spanish wool, became a magnet for intellectuals and dissidents of all persuasions, blossoming into a center for the Continent's intellectual avant-garde.

The university actively recruited great scholars in an international acquisition drive. The legendary Herman Boerhaave taught medicine and anatomy and staged public dissections in the name of science; the boy genius Hugo Grotius, one of the fathers of international law, studied here with the French Protestant and humanist Joseph Scalinger (1540–1609), who was, among other things, an authority on Arabic grammar who advocated the study of the Quran and its language as purely scientific subjects. Carolus Clusius,

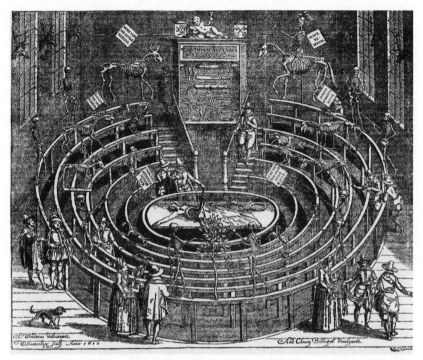

Knowledge gained by scalpel. Leiden University was one of Europe's
intellectual and academic centers.

formerly botanist to his imperial majesty Rudolf II, was enlisted
during this recruitment campaign.

The expanding university changed life for everyone in Leiden.
Students and scholars needed boardinghouses, servants, amuse-
ments, libraries, bookshops, taverns. Moreover, the city's reputation
as a place of tolerance and learning with access to excellent libraries
attracted others on the university's periphery: Baruch de Spinoza
and René Descartes sought refuge here, as did many freethinkers
and dissident philosophers. Different religious traditions and philosophical perspectives converged here—not only in the library, but
also at the local watering holes.

Fleeing religious prosecution, the printer and publisher Christoffel Plantijn set up shop in Leiden in 1576, the year after the foundation of the university. His establishment in the town was the
beginning of a flourishing publishing industry specializing in Asian

languages and non-European scripts. In Leiden, authors could find printers and editors who would be at home not only with Greek and Hebrew letters—as would be expected of any university teaching biblical philology—but also with Arabic movable type, as well as Chinese characters.

The rebirth of Leiden as a university city was part of a larger cultural and economic awakening in the Netherlands. Until the mid-sixteenth century, the seven provinces had amounted to little more than a landscape of farmers and herring fishers, with the odd town or commercial harbor dotted along the flat horizon. It was a poor country without large cities, courtly culture, or powerful aristocratic dynasties. For centuries, the cultural, intellectual, and economic center of the region had been to the south, in splendid and wealthy Bruges, in Ghent, and especially in Antwerp, which had developed into the main hub of European seaborne trade north of the Alps.

During the Eighty Years' War, parts of the Netherlands rebelled against their Spanish and strictly Catholic overlords, who ruled the land with an iron fist, brutally suppressed all Protestant leanings, and funneled off vast amounts of taxes to Madrid. In spite of all this, King Philip II found himself bankrupt, unable even to pay his troops. In 1576, soldiers in his army who had not received pay for months decided to take matters into their own hands. Six thousand mutinous troops attacked the city of Antwerp, whose Spanish commander did not intervene. In a fury of rape, pillage, fire, and murder, the soldiers slaughtered some ten thousand inhabitants and incinerated eight hundred houses in the center of the city. Among the many survivors who fled this atrocity was the printer Christoffel Plantijn.

Also known as the Spanish Fury, the sack of Antwerp would be a milestone in a saga of gradual decline. Amsterdam now became the rising star on the international market. Around 1500, the little town built around the dam on the River Amstel had counted some twelve thousand inhabitants. By 1620, almost ten times as many people lived in the bustling, bursting, and increasingly beautiful city whose harbor was a true gate to the world.

Having built its wealth and trading networks on the reliable staple of Baltic grain, Amsterdam benefited greatly from the agricultural crisis, using this chance to realize ambitions that went much, much farther. While grain was dependable, the truly exciting possibilities and the greatest profits lay in the overseas trade in spices, sugar, porcelain, tobacco, tea, and coffee—luxury products for which a growing number of urban professionals were willing to pay a pretty penny.

The can-do attitude and the self-confidence with which the Dutch elites grasped this historic opportunity for transformation is astonishing. "God created the world, and we created the Netherlands"—this turn of phrase proudly made the rounds among the Dutch who were indeed remaking their own world, almost down to its Creation. Even the pitiful size of the Dutch territories, the country's traditional problem that was surely based on God's will, now became subject to the will of administrators, engineers, and entrepreneurs. Vast networks of canals were dug in the marshlands and countless windmills were erected to pump the water out to sea. Even more spectacular were the polders, dikes, and breakwaters constructed to reclaim land from the sea itself—long, still-barren stretches of sandy plains over which the linen sails of the windmills were racing day and night, land for breeding cows and growing vegetables, land that was a human Creation.

At the Beemster Polder in the west of the country, forty-three mills were draining the marshes between 1607 and 1612, freeing up some twenty-seven square miles (seventy square kilometers) of new land. Roughly a sixth of the original surface of the country was added not by waging war but by reclaiming land from nature. Tens of thousands of trees shipped from the Baltic were rammed into the marshy soil around old Amsterdam to create a magnificent semicircle of grand canals. Bordering these canals were handsome houses whose playful compromises between Calvinism and cheerful wealth proclaimed the city's success to the world. Unsurprisingly, Dutch engineers were soon recognized internationally as specialists in coastal construction—so much so that they were also involved in the construction of St. Petersburg and in large land-reclamation projects on the east coast of England.

The dynamic transformation of the Dutch intellectual, economic, and even physical landscapes demanded that everyone submit to the brisk pace of historic change—a difficult and precarious process for many people. The cheap grain being unloaded onto the docks of Amsterdam dictated prices throughout the land, as the Little Ice Age made itself felt with even greater severity in the Northern European countries, where conditions for agriculture had always been rather marginal. Many farmers, and especially many young people who either were not going to inherit land or could no longer make their land yield enough to feed themselves and their families, moved into the cities, where they could work in domestic service, on the docks, or in other menial occupations.

Many of those who remained in their villages heeded an early form of mass media by following advice they found in almanacs, popular calendars, and yearbooks containing information on saints' days and Bible quotes, as well as illustrations and words of wisdom and advice. The compilers of these almanacs were aware of the then-fashionable wave of publications on natural sciences, botany, and plant and animal breeding, and they shared aspects of this knowledge with their rural readers.

Forward-looking farmers now began to put their energy not into producing grain, which was proving unprofitable against imported competition, but into vegetables for the hungry city, as well as cattle for milk, cheese, and meat—cattle that could graze peacefully on the grassy meadows of reclaimed land. They also overcame the problem of fallow fields by sowing clover as winter feed for the cattle. It is now understood that clover enriches the soil with nitrogen, crucial for plant growth, preparing the field for different produce. By becoming mainly producers of items for urban markets instead of for their own needs, these farmers were already working and thinking in modern ways.

THE CITY DEVOURS ITS CHILDREN

The exodus from the countryside was a European phenomenon, as the crisis of rural living deepened and grain prices rose ever higher. But while Amsterdam grew, most urban areas failed to expand, despite the continuing influx of migrants. The explanation of this seeming paradox lies in the living conditions the newcomers faced. Work was often hard and paid a pittance, accommodations were cramped and shared with many others. As a result, immigrants married later (domestic servants were often prohibited from marrying at all) and had fewer children. They were also the first to be victims of epidemics, starvation, work-related accidents, random violence, or simply exhaustion. Since their occupations were risky, their lives were shorter than those of better-established city dwellers.

Those seeking success in the city needed to have qualities that country folk lacked. In the city, after all, you could and had to reinvent yourself. A medieval town, with its guilds, price controls, and social rigidity, offered nothing more than menial jobs. A modern city, however, was a quite different creature. Here, the Netherlands and England were ahead of most of their European neighbors. They made it relatively simple for educated and entrepreneurial newcomers to succeed and to build new positions for themselves.

The life of painter Rembrandt van Rijn (1606–1669) exemplifies this new social mobility and transparency. Rembrandt was the son of a Leiden miller, who was himself the son of a miller. Had he been born a generation earlier, the boy undoubtedly would have entered the family trade. Instead, his father decided to send him to Latin school, making him the first child in his family to receive a formal education and giving his career a completely different trajectory. After breaking off his studies at Leiden University, he was apprenticed first to the painter Jacob Isaacszoon van Swanenburgh and then to Pieter Lastman in Amsterdam.

When Rembrandt struck out on his own, he began to make a very comfortable living as a portraitist of those who had "made it" in Amsterdam and were ready to convert hard-earned cash into social

The impertinent gaze: Rembrandt, a miller's son, was part
of a newly assertive middle class.

prestige. He remained fascinated by the world of classical antiquity, which he had first encountered at school, and by the vast stream of exotic goods, objects, and stories arriving with the trading ships from across the seas. He himself became a collector, an art dealer, a man of learning, an artist deeply curious and continuously seeking to find new modes of expression. He made use of objects from other cultures not only as props for canvases but also as modes of inquiry into the art of painting itself. His life had moved a long way from the lives of his father and grandfather.

Word of unknown civilizations and natural wonders from far beyond the physical horizon entered Amsterdam through its teeming port, changing not only Rembrandt's outlook on the world but also the city itself.

Not since ancient Rome or even Constantinople had so much diversity, so many languages, nationalities, cultures, and wares rubbed up against one another as in seventeenth-century Amsterdam. Exiles, rebels, and religious dissidents from all over Europe met in alehouses and in learned societies to discuss the newest publications flying off the printers' presses in a variety of languages—often for clandestine export into France or other, more repressive European countries.

Scientists, scholars, and speculators crowded the harbor to scour ships' cargoes and seamen's boxes for unusual merchandise, mute witnesses of other civilizations, and specimens of unknown animals and plants to study, discuss, and explicate. The reception rooms of wealthy burghers were decorated with Oriental carpets, Chinese porcelain, and mahogany furniture. Their owners learned to drink tea and chocolate, gentlemen and sailors alike smoked clay pipes, and the wealthy grew to appreciate fine wines and to use cane sugar in their desserts. A global trading empire came right into their homes.

Like most of his educated contemporaries, Rembrandt was fascinated by the vastness of the world and the strangeness of the exotic objects reaching the city. Some of these objects suggested that there were startling human similarities beneath the manifestations of cultural differences. The painter created several self-portraits in which he wears Oriental robes, and he was eager to meet those who brought aspects of a different life to the city. He painted dignitaries of the city's Jewish community, largely consisting of families who had been forced to flee from religious persecution in Portugal. One of his subjects was the merchant Michael d'Espinoza, whose son Bento (Hebrew: Baruch) will appear in a later chapter.

This new world held immense promise, not least of fabulous profits to be made by importing the riches of other continents. But huge opportunities also entailed huge risks, particularly when it came to overseas trade. A single shipload of Oriental wares could make a merchant rich overnight, but manning and equipping a ship (which had to be leased or bought), including sailors' provisions for several months as well as the merchandise itself and weapons to defend it, was vastly expensive. A vessel trading with the West Indies, or with Indonesian, Chinese, South American, or African ports, would be gone for months, and ship and crew would have to contend with all manner of dangers, including storms, scurvy, piracy, mutinies, corrupt harbor officials, ill winds, and the quiet danger of shipworms eating through the ship's hull. This was a perilous life, and a ship failing to return could spell not only death for its sailors but also ruin for the merchant who had sent it out.

The solution for this problem originally came from Italy, but in Amsterdam it was realized on its greatest scale. Merchants lowered the financial risk of individual journeys by selling shares in the projected profits, inviting others to share the burden, and potentially the profits. Everyone from vastly wealthy patricians to elderly widows who had no more than a few spare guilders invested in this scheme, which began operating in 1602 and came to be known as the *beurs*, or the stock exchange. The mighty Vereenigde Oostindische Compagnie (VOC), the main actor, sent hundreds of ships out into the world. Investors were so enthusiastic that when the stock exchange was set up, 1,143 individuals created a trading capital of 3,679,915 guilders, the equivalent of one hundred fifty million American dollars today.

But even in canny Amsterdam, not all investments were necessarily based on solid business practice. Indeed, investment opportunities became so sought-after that even the work of the quiet scholar Carolus Clusius in his botanical garden was swept up in the furore. The tulips that he had introduced to the Netherlands and bred in many permutations became the focus of the first modern stock-market bubble when investors clamored to get their hands on the bulbs. It had become fashionable for the rich to display rare and exotic tulips in their houses, and the bulbs proved ideal for speculation and resale.

From 1630 onward, prices for certain varietals of tulip spiraled up and up, until a single bulb could command an astronomical price. The most precious varietal, *Semper augustus*, could cost as much as a well-appointed country house. This tulip mania inspired breathless buying and selling until the bubble burst in February 1637, leaving many investors destitute and driving some to suicide. Having been traded for vast sums on one day, the worthless bulbs were suited only for the garden in the days that followed. At least their blossoms were as colorful as they had ever been.

THE MAGIC OF GREEN CHEESE

During the late sixteenth and the seventeenth centuries, the Netherlands became a kind of laboratory for a new way of living and of seeing the world, a perspective that had little in common with that within feudal or late medieval societies. Its driving force was a remarkably rapid changeover to a society centered on a market, rather than a fortress. This vast change led to considerable social unrest and violent resistance, particularly among poorer Europeans. Rural areas were especially affected by waves of violence as the poor demonstrated against rising prices and against the limitation or abolition of their ancient, customary rights in favor of market efficiency.

One of these rights, which had been customary in much of Europe and particularly in England, was the institution of the commons. Every medieval village had a common (or commons), a piece of land where everyone was allowed to graze cattle and cut winter feed. This was crucial to the survival of the landless poor, who could use the commons to raise a goat, a cow, or a few chickens and therefore have access to milk, eggs, and occasionally meat or the revenues from selling their livestock.

Legally, the commons were in uncertain territory. Like almost all land, they nominally belonged to the landowner, whether an aristocratic family or the Church. In practice, however, a habitual-usage right was respected, and the owners did not use this plot of land for their own profit. Already in the fifteenth century, there had been tensions between peasants and landowners who found that their land could yield more if put to different uses, particularly for breeding sheep, since the animals required a minimum of human intervention to raise, and wool cloth was easy to export.

During the second half of the sixteenth century, these conflicts intensified in England. The Crown often sided with the country folk against the landowners, not only to protect the peace but also in a bid to keep the latter from becoming too powerful. Even royal intervention, though, could not prevent the situation from escalating.

As the landowners experienced the decline in revenues caused by increasing bad weather, they began to look for alternative income streams. They saw that the commons would be ideal for wool production, even if that meant moving the poorer folk off the land. English textile mills were willing to pay good money for good wool, as English cloth was beginning to outcompete Spanish, Italian, and Flemish products. Fencing in the commons and forcing the poor to relinquish its use marked the beginning of an economic renaissance.

The matter looked somewhat different from the perspective of the evicted people. They had always lived off the commons, and many had no other means of supporting themselves and feeding their families. Without access to some land, even common land used by many people, they were unable to survive in the country. Destitute, unskilled, and hungry, they were now forced to look for work in the cities, in the army, or in the spinning and weaving concerns that were also buying the wool from sheep raised on their former commons. Tens of thousands of peasants were uprooted, with no choice but to leave their homes in this new process of exploitation and profiteering. Similar instances of increasing wealth accumulation in the hands of a few had been described by Thomas More as early as 1514. Now, with the enclosure not only of agricultural fields but even of the common land itself, it was becoming a new norm:

> Your sheep [. . .] that commonly are so meek and so little, now, as I hear, they have become so greedy and fierce that they devour men themselves. They devastate and depopulate fields, houses and towns. For in whatever parts of the land sheep yield the finest and thus the most expensive wool, there the nobility and gentry, yes, and even some abbots though otherwise holy men, are not content with the old rents that the land yielded to their predecessors. Living in idleness and luxury without doing society

any good no longer satisfies them; they have to do positive evil. For they leave no land free for the plough: they enclose every acre for pasture; they destroy houses and abolish towns, keeping only the churches—and those for sheep-barns.[3]

This "positive evil" was regularly opposed by desperate smallholders seeking to preserve their livelihoods. In 1607, rebels assembled in Northamptonshire, Warwickshire, and Leicestershire to protest the ongoing enclosures of common land. Their commander was a tinker from Northamptonshire named John Reynold, known by his supporters as Captain Pouch. The nickname came from the small leather pouch, carried on his belt, that, so he claimed, lent him magic powers and protected all his followers from harm. His mission was, he said, blessed by God and by the king. Soon, at Reynold's instigation, some eight thousand men, women, and children were busy ripping out border hedges, filling in ditches, and razing the new fences enclosing the commons.

Faced with this sudden rebellion, the city fathers in Leicester erected a gallows as a warning to the rebels, who immediately tore it down. The town militia refused to obey the order to disperse the crowds and crush the unrest. Only when the landowners, including some powerful nobles, mobilized together, arming their own men, did they succeed in overwhelming the rebels and reimposing their authority. The demands of the people went unheard. Instead, their leaders were arrested, hanged, and quartered. One of these unfortunates was Captain Pouch himself, whose leather bag, however, remained a focus of tremendous interest. A witness described what was found inside: "when hee was apprehended his Powch was seearched and therein was onely a peece of greene cheese."[4]

Even though Reynold's cheese had failed to work its magic and save him from the gallows, the rebellions continued, especially because the enclosures of the commons resulted in even higher grain prices, since less land was available for grain production. Violent resistance flickered into life not only after particularly bad harvests or large price hikes, but in many cases when public goods were seen to be effectively privatized—as was the case in the Forest of Dean and Gillingham Forest between 1628 and 1631.

While agriculture was in crisis due to failing harvests, the reorganization of land use and ownership in large parts of Europe was further changing social structures and power relationships. From Poland to Italy and the Netherlands, farms and estates were moving away from subsistence toward production for local or even international markets. Measures such as the enclosure of the commons consolidated land use in fewer hands and swelled the ranks of the landless poor.

Often these changes met with violent resistance. England was not the only country where rebellions were caused by enclosures, the abolition of long-held rights and privileges, or higher taxes. In France, the Croquants in the Perigord in 1594, 1624, and 1637, and the revolt of the Va-nu-pieds in Normandy in 1639, to name only the two largest, were among an estimated 450 local armed uprisings documented between 1600 and 1715.

In Spanish-ruled southern Italy, and particularly in Naples, bread riots occasionally led to great excesses of violence; Catalonia was the scene of many rebellions, as was the rest of Spain. Central Europe, meanwhile, had sunk into the murderous chaos that was the Thirty Years' War, which would cost the lives of a third to half of the people in the affected areas. In Russia and the Ukraine, almost half of the population had perished during the Time of Troubles.

The violent restructuring of rural economies in Europe, in tandem with the effects of the Little Ice Age, resulted in the uprooting and expulsion of hundreds of thousands of smallholders, landless tenants, and farmworkers who were forced to give up their homes and their way of life. The lands they vacated were made more profitable by being frequently leased to commercial operators, while the management of country estates and large farms was now more often entrusted to professional farm managers who knew how to make a profit out of the land. For these large concerns, the uncertain harvests of the era were less of a problem than they were for the peasants, for the wealthy no longer immediately depended on the harvest for survival. For them, grain had simply become one more commodity to be traded; moreover, bad harvests meant higher prices.

THE GREAT TRANSFORMATION

The enclosure of the commons was one of the key moments of economic and political change highlighted by the Austro-Hungarian economic historian Karl Polanyi in his insightful *The Great Transformation* (1944). More than most of his peers, Polanyi acknowledged and analyzed the powerful connections between the way societies do business and the practices and beliefs they follow. The story of the rise of Amsterdam illustrates how a society created a strong market that in turn deeply changed the society through new wealth and new poverty, along with new systems of public education and laws and regulations. Also new were such social practices as pragmatic tolerance toward strangers and the toleration of intellectuals and book dealers; their business may have been politically dubious or religiously unacceptable to some, but it was also very lucrative. Amsterdam and cities like it did not just become richer—they became different.

Polanyi reached this conclusion by precisely inverting the tradition of economic analysis that had started with Adam Smith and gained its greatest influence under two men of Polanyi's generation, fellow Austrians Friedrich von Hayek and Ludwig von Mises, who tended to describe the economy as a self-contained system animated by the eternal human search for profit: "The outstanding discovery of recent historical and anthropological research," argued Polanyi, "is that man's economy, as a rule, is submerged in his social relationships. He does not act so as to safeguard his individual interest in the possession of material goods; he acts so as to safeguard his social standing, his social claims, his social assets."[5]

Viewing the economy not as a closed system apart from the rest of human life, but rather as one aspect of more complex social activities that powerfully influences the others, turns out to be immensely fruitful for understanding the profound changes in all areas of life that occurred during the height of the Little Ice Age.

Since 1944, when Polanyi's study was first published, historical horizons as well as available data have changed. But while some historical examples might be seen differently today, the thrust of his argument remains valid. It runs as follows: In premodern, feudal societies, the goal of economic activity was directed not toward wealth accumulation and social climbing, but toward maintaining status in a rigid social hierarchy in which every individual had a more or less clearly defined role and in which social capital—prestige, honor, respect—was deemed more important than financial capital. The "great transformation" changed that. It was an economic revolution, but its implications affected all aspects of life.

Before the advent of market economies, peasants made up the majority of Europe's populations—born into their farms or villages and into a social hierarchy, a particular faith, and very often a traditional occupation. They tended to stay where they were born unless wars, epidemics, or famines forced them to leave. Farming methods had remained "largely unchanged" since antiquity. After all, agricultural workers had no incentive whatsoever to increase their productivity, as everything beyond what they needed to survive and to sow for the following harvest was taken away by their lords. The more they worked and produced, the more would be taken from them. Once famine was kept at bay, there was no point in working more—especially in societies in which work was physically hard, and regarded as a burden to be endured with Christian patience.

In medieval cities, markets and economic practices were very strongly embedded in their social contexts. Prices were strictly regulated, competition was limited, products and methods of production were dictated. A craftsman wanting to set up shop had to be part of a guild that controlled not only his skills and the quality of his products but also how he lived his life, how often he attended church, and whom his children might marry. The guild and its rituals, its festivals and assemblies, religious processions, masses, and occasionally even special clothing dictated not only economic matters but also social activities and identities.

While aristocrats and the princes of the Church (most of whom were themselves from aristocratic families) were clearly in very privileged positions, they also had to obey a set of rigid rules, or at least to give the appearance of doing so. They alone could own land, which they could "liege" or lease to others in return for taxes, labor, or armed service. The rules of their class prevented them from working with their hands or going into business. In some countries, such as France, it was actually against the law for nobles to be in trade—which did not prevent them from profiting handsomely from trade being carried on by others on their behalf. Even so, they were not allowed to sell their land. Land remained *extra commercium*—exempt from commerce. This "real estate," the cornerstone of the feudal order and hierarchy, was not for sale, and markets had only a limited geographical and social reach. "Gain and profit made on exchange," noted Polanyi, "never before played an important part in human economy. Though the institution of the market was fairly common since the later Stone Age, its role was no more than incidental to economic life."[6]

Polanyi described the far-reaching transformation of this system of estates during the sixteenth and seventeenth centuries through the rise of markets, though he does not factor in the Little Ice Age, which, when he was writing in the early 1940s, was very little known and certainly little researched. Even so, he was struck by the sheer speed of the rise of markets as principal agents of economic and social change: "The transformation to this system from the earlier economy is so complete that it resembles more the metamorphosis of the caterpillar than any alteration that can be expressed in terms of growth and development."[7]

Although this process unfolded at different speeds and with different effects in different European countries—Renaissance Italy and later the Netherlands and England underwent these processes much earlier than Spain or Russia, for example—it was deeply pervasive. Even if the representatives of the old order—guilds, corporations, aristocrats—resisted this change, they eventually had to accept it, along with the social transformation accompanying it:

> Deliberate action of the state . . . foisted the mercantile system on the fiercely protectionist towns and principalities. Mercantilism destroyed the outworn particularism of local and intermunicipal trading by breaking down the barriers separating these two types of noncompetitive commerce and thus clearing the way for a national market which increasingly ignored the distinction between town and countryside as well as that between the various towns and provinces.[8]

As with the enclosures of the commons and the employment of professional estate managers whose task it was to realize profits, as with the powerful East and West India Companies in Amsterdam and London (and elsewhere)—which were run to generate revenue and financed by stock exchanges—the upswing of economic activity changed not only the practices of the markets but also the mentality of the people participating in them. Mercantile, profit-oriented practices began to outweigh those geared toward social cohesion and the maintenance of status. Long-distance trade, according to Polanyi, perfectly illustrated this point:

> As to food supplies, regulation involved the application of such methods as enforced publicity of transactions and exclusion of middlemen, in order to control trade and provide against high prices. But such regulation was effective only in respect to trade carried on between the town and its immediate surroundings. In respect to long-distance trade the position was entirely different. Spices, salted fish, or wine had to be transported from a long distance and were thus the domain of the foreign merchant and his capitalistic wholesale trade methods. This type of trade escaped local regulation and all that could be done was to exclude it as far as possible from the local market.[9]

Since long-distance trade did not have to abide by local rules, it had a tendency to upset traditional hierarchies. The catastrophic harvests of the 1590s and the population movement away from the country and into the cities strengthened trading links and markets. Cities, after all, could not feed themselves using local produce alone. They depended on imports, and when the granaries were empty, even cities in Spain and southern Italy were forced to buy their foodstuffs from as far away as Amsterdam.

Polanyi's argument can readily be enlarged to take into account what we now know about the Little Ice Age: The social and economic system of feudal European societies rested on landownership and local grain production. This was its central pillar as well as its main vulnerability. When temperatures declined enough to disturb grain production and therefore undermine this pillar, the entire social model fell into decline. Europeans were forced to think of alternative ways of organizing themselves and their economic life. A way of life that had been relatively stable for a millennium was tottering.

When the effects of global cooling became obvious, many contemporaries interpreted them as divine wrath or the result of witchcraft. Time and again, women and men were burned at the stake, time and again priests and monks attempted to reestablish the natural order using relics, processions, or holy water. Especially along the Mediterranean coast, such rituals were common occurrences. At the same time, though at very different rates of speed and penetration into societies, different kinds of knowledge and different perspectives on the forces of nature were gaining ground. In agriculture, these changes proved to be particularly important.

As we have seen, Dutch farmers had already been forced to make significant changes toward the end of the sixteenth century. In response to the influx of cheap Baltic grain and the worsening climate conditions, they moved into beef and milk production, selling milk, cheese, butter, and meat, as well as vegetables and even flowers from their market gardens, and experimenting with new crops such as potatoes. They soon found, to their surprise, that with more diverse crops and more manure to fertilize their fields, even grain production began to provide better yields, which made farming more resilient in difficult conditions.

This kind of knowledge spread fast, carried by people hired for the purpose, people fleeing from wars, and increasingly also by agricultural books, tracts, almanacs, and letters that were translated and read in several languages. Protestant farmers from the Netherlands, for instance, fled to southern England in order to escape the cruel and seemingly endless wars of their own region against Spain. In their new country, they sought employment on farms, and local farmers then adopted some of their methods.

A more momentous change, however, came from printed works describing new agricultural techniques. Books such as *Rei Rusticae Libri Quatuor* by Conrad Heresbach (1570), Martin Grosser's widely read *Kurze Anleitung zu der Landwirtschaft* (1590), Johann Coler's annual *Landwirtschaftlicher Kalendar* (first edition 1590), and the dramatically entitled *Le Théâtre d'Agriculture* by Olivier de Serres (1600) went through several editions and introduced a new empirical and methodical approach to farming. The results were there for all to see, and to eat: While medieval farming techniques had rarely yielded more than four grains of cereal per grain sowed, modern farming methods in the Netherlands, southern England, and parts of France could already harvest six or seven grains. As natural conditions worsened, technical and cultural adaptations were literally beginning to bear fruit.

Another important element enabling European populations to grow again was the "discovery" of other continents and their colonization. The first powerful factor was the institution of a system of methodical and often brutal economic exploitation, which would soon flood the coffers of colonial rulers with money. The second factor was the "Columbian Exchange," as Alfred W. Crosby called it, the witting or unwitting swap of biological materials—from viruses to plants to domestic animals—between the colonizers and the colonized.

For the indigenous peoples of the Americas, the biological consequences of the shipborne invasion of their continent ranged from the introduction of dogs, pigs, and horses to the outbreak of hitherto-unknown diseases such as smallpox, tuberculosis, bubonic plague, and cholera, which in many cases all but destroyed populations lacking resistance to these diseases. In return, the Europeans received not only syphilis (hailed by churchmen of all stripes as God's punishment for immorality) but also turkeys and many of the plants that were first cultivated so carefully by Carolus Clusius and his colleagues in European botanical gardens—including tomatoes, pumpkins, sweet corn, tobacco, and especially potatoes.

At the end of the sixteenth century, these bounteous tubers were still grown mainly in botanical collections and were eaten only in the French regions of Franche-Comté, Vosges, Lorraine, and Alsace, as well as in parts of the Low Countries. The rest of Europe was still waiting for this precious staple food. In Central Europe and England, the commercial growing of potatoes would occur only several decades later.

Potatoes had great advantages, but they also had a powerful foe: the instinctive conservatism of rural populations. The first German-language cookbook containing recipes for the preparation of the strange tubers was written in the Seitenstetten Monastery in Austria and bears the date 1621. Perhaps the Thirty Years' War contributed to the spread of this strange-looking vegetable. Most armies had only weak supply chains and lived off the land—i.e., by robbing the local populations and impounding their supplies. In order to make this more difficult for an advancing enemy, most commanders relied on a scorched-earth policy. Entire fields of ripe grain were torched, but it is much more difficult to destroy a plot of potatoes. The plants will not burn, so they have to be ripped out individually. Those who planted potatoes had a little more food security than those who relied only on grain.

Although rural people generally regarded the novel tubers from overseas with intense suspicion, the various names given to the potatoes reveal the ways in which they were circulated and made acceptable. First grown in botanical gardens, where they were planted for their pretty and abundant flowers, they were introduced as food first in Flanders, northern Italy, and Brandenburg, all regions severely affected by frequent grain shortages. The Italians first named the strange vegetable *tartuffo,* meaning "growing in the ground." The scholar Serres's *Theatre of Agriculture* introduced them to France as *cartoufle,* but they had great difficulty getting established there, and they were eventually known as "earth apples," *pommes de terre.* (Another recent arrival from the New World, the tomato, was engagingly called *pomme d'amour.*) The French word *cartoufle* would later evolve into the German *Kartoffel,* but this happened only after potatoes were already well established in Prussia, where their cultivation was strongly encouraged by the government. Potatoes were ideal for the sandy soils of Brandenburg, and they quickly became a staple. From here they spread southward, to the Czech-speaking Habsburg lands, where they became known as *brambor* (from Brandenburg). Even farther south, in Romania, they were called *brandaburca.* The potato's entry onto the culinary stage as a staple diet for the poor would create a long-lived prejudice against tubers, especially in France. In 1765, more than a century after the first potatoes had been harvested in France, one author archly wrote in Diderot and d'Alembert's great *Encyclopédie*: "One justly reproaches the potato for causing wind; but of what importance is that for the robust organs of farmers and day labourers?"[10]

New crops and new techniques enabled European farmers to cope with the worst consequences of the Little Ice Age by diversification and by producing for the market. This change of methods and approach, however, had far-reaching consequences. It dislocated countless smallholders and agricultural workers, swelling the ranks of the landless poor. Most of these, or their descendants, would eventually find employment of a harsher kind in the factories and textile mills from which the Industrial Revolution would grow. Farms had become consolidated, reorganized to deliver produce to urban markets and to generate revenue for their owners, rather than subsistence to those who worked them.

The result was, as Karl Polanyi writes about the enclosures of the commons in England, a "revolution of the rich against the poor." Traditional ways of life and traditional social models, already shaken by bad weather and failing harvests, were coming under an altogether new pressure.

A PICTURE OF THE WORLD

The gradual transformation of farming practices by early agricultural scientists and profit-seeking estate managers, the rise of long-distance trade, and the increasingly professionalized administration of taxes, land, and state bureaucracies were creating the foundations for a new Europe. But it would have been impossible without another development: From 1600 onward, throughout Europe more children (mostly boys) were going to school—learning reading, writing, arithmetic, and a bit of the classics, as well as a smattering of geography and history. Many urban people, and even some in the countryside, were now receiving some formal education.

For centuries, education had been almost exclusively the domain of the Church, which had been careful to use this great privilege to maintain its power and prevent those in its care from religious doubt. But the Reformation destroyed this monopoly and created, in educational terms, a competitive dynamic. For Protestants, the ability to read the Bible and enter into a direct conversation with the Creator was an essential part of religious practice. Protestant rulers and clerics had founded schools to teach even the children of simple folk; it was all a part of bringing up good Christians.

The Catholic Church was forced to react to this threat from Protestant literacy, which had proven capable of attracting many people and even entire countries from the old faith by challenging priests who were largely ignorant of the Bible and, where necessary, by quoting chapter and verse in fiery sermons. The Catholic Counter-Reformation was an attempt to repress and if possible outroot the influence of Martin Luther, John Calvin, and other Protestant leaders, not only by military means but also through a massive cultural offensive. A major part of this effort was the establishment of Catholic schools, where children would be drilled in the Catholic

Catechism and thereby supposedly become inoculated against Protestant heresy.

The Jesuits, founded in Spain in 1534 as an order of militant intellectuals, played a key role in this ambitious project. Their founder, Ignatius of Loyola, had been a soldier who had "seen the light" while recuperating from a battle wound. He sent his *Milites Christi* (Soldiers of Christ) not just throughout Europe but far around the globe, on missions to spread the faith among those he viewed as heathen peoples. Some of these Jesuits, like the great scholar Matteo Ricci, even worked as far away as the imperial court of China.

In Europe, the Counter-Reformation created a new role for the Jesuit order, which soon began to concentrate on education as a means of reestablishing the dominance of the Catholic Church. By 1574, Jesuits were running 125 schools throughout Europe; by 1608, the number had swollen to 306, and a decade later to 372. By the mid-seventeenth century, there were more than five hundred Jesuit schools across Europe and around the world. In France, Jesuits led not only the prestigious college in La Flèche in which twelve hundred children of the social elite were educated, but also Douai University in the north of the country, an important education center not only for young Frenchmen but also for clandestine English Catholics who were now being persecuted in their Protestant homeland.

The influence of Jesuit education was profound, and it often worked in unexpected ways. The literary historian Robert Wilson has argued intriguingly that it is possible that the young William Shakespeare was part of this Catholic underground (his father was a recusant Catholic) and that he may have spent some time in secret Jesuit establishments in Yorkshire, and then in Douai. While there is no incontrovertible evidence for Shakespeare's being a Catholic, Wilson notes that classical allusions in many of his plays seem to emanate straight from the Jesuit syllabuses of the time. He also notes that the Jesuits were noted for using novel pedagogical ideas, and that the staging of theater productions by pupils had a special place in their concept of education. Imagining England's bard as a clandestine papist in his youth is a daring change of perspective, and it makes many of his literary and personal preoccupations appear in a different light—not least his fascination with hidden identities, his legendary concern for his own privacy, and his manifest pity for those fleeing religious persecution.

It was not only the Jesuits who hoped to transform society and its ills (religious or otherwise) through an education that was more adapted

C.
Inſtrumenta Muſica.   Klangſpiele.

Muſicalnſtrumenta ſunt, quæ edunt Vocem:	Klangſpiele ſind / die eine Stimme von ſich gebe:
Primò,	Erſtlich /
cùm pulſantur,	wann ſie geſchlagen werden /
ut *Cymbalum* 1	als / die Cymbel 1
piſtillo,	mit dem Schwengel ;
Tintinnabulum 2	die Schelle / 2 (Knöpflein;
intus globulo ferreo;	inwendig mit einem eiſernen
Crepitaculum, 3	das Klepperlein / 3
circumverſando;	durch umdrehen ;
Crembalum, 4	die Maultrommel / 4
ori admotum,	an den Mund gehalten /
digito :	mit dem Finger ;
Tympanum 5	die Trummel 5
& *Ahenum*, 6	und Pauke / 6
Claviculâ, 7	mit dem Schlegel / 7

Learning through pictures: the schoolbook *Orbis Pictus* taught generations of children to read, write, and understand the world.

to young minds and interests than the classical methods of rote learning and regular whipping. The Protestant theologian and teacher Jan Amos Komensky, better known by his latinized name of Comenius (1592–1670), was bishop of Sárospatak in today's Hungary; here he experimented with new ways to enlist the curiosity of children to make them better learners. The result of his efforts was published in 1658 under the title *Orbis Sensualium Pictus* (The Visible World in Pictures).

Orbis Pictus—as it is often called—was revolutionary because it was based on the concepts and interests the pupils might already possess when they arrived at school. Illustrated with woodcuts, simple German and Latin sentences explained the everyday world in which the children were living: *ovis balat*—the sheep bleats; *infans ejulat*—the child cries; *mus mintrit*—the mouse squeaks. There were sections about nature, trades and crafts, human anatomy, and even other religions, which were described not just as damnable heresies but also as different ways of life. There were sections on sciences and, finally, in the last chapter, came a preview of the Last Judgment. It was a comprehensive compendium of an early modern world in simple language, and it proved so successful that the first English translation was published just one year after the original edition. Reprints were still coming off the presses until well into the nineteenth century. Arguably, Comenius, a provincial Hungarian theologian, influenced more European children over a longer period than any other educator.

Going to school, or, for wealthier children, being taught by a private tutor, became a comparatively common experience in Europe around 1600. While public schools were almost entirely reserved for boys, girls were also learning to read and write, albeit generally at home.

While a great increase in printed materials is clear evidence, it is remarkably difficult to determine the degree of literacy among early modern Europeans. The vast majority of people simply did not leave behind personal written documents such as diaries or letters, and relatively few had books, which were quite expensive. In Protestant households, one might have found a Bible, but that in itself was not proof of how many people in the household could read from it, how many could read fluently, and how many could not only read but also write—especially since reading and writing were often taught as two quite separate skills. One approach to estimating the degree of literacy has been to look at signatures on contracts, marriage registers, ships' lists, and other documents. Signatures, however, were frequently memorized, and they often seem to have been the only thing a person could write, so they provide little indication of actual functional literacy.

Another difficulty lies in the great regional, social, and religious differences among levels of education, which are notoriously difficult to establish or compare in any detailed manner. City dwellers and wealthy burghers tended to be more educated; Northern Europeans were more likely to have gone to school, at least for a few years, than people living south of the Alps; men were more likely to be literate than women, and Protestants more than Catholics. Different rulers showed different attitudes toward educating their subjects; progress and regression alternated. In spite of all this, however, it is clear that the educational landscape of Europeans was changing fundamentally. Historians speak of an education revolution that began in England and the Netherlands and moved to Germany and France, while never quite taking hold in Eastern or Southern Europe.

Almost everywhere, the vast majority of peasants and day laborers were illiterate. Interestingly, the same tended to be true at the upper end of the social pyramid: Aristocrats were often resolutely uninterested in educating their children beyond the absolute basics. Their role and power depended on birth, not education. They needed to be able to hunt and command, to dance and have children. For written business, there were always secretaries.

In the middle range of society, a very different dynamic was driving much of the overall change. Merchants, lawyers, teachers, administrators, and managers, as well as such specialized professions as goldsmiths, apothecaries, and ships' captains, were mostly literate by 1600; only fifty years earlier, roughly half of them had not been able to read or write. Among craftsmen and workers in Protestant countries, literacy rose to about fifty percent; during the same period, the proportion of university graduates in the Netherlands rose from one to three percent of the population.

While we cannot know exactly how many children attended schools, precisely what they were learning (in many cases, the teachers themselves could barely read or write), and how fluently they were able to read and write when they left, it is easier to reconstruct how much people read. Inventories of estates in Kent in England, for instance, show that around 1560 only about ten percent of wills included books, while half of all households had at least some books in 1620. In northern Italy, around one percent of inventories itemized a library of more than a hundred volumes, while this number rose to ten percent around the year 1600.

Private book collections, though, do not paint a full picture. Learned publications were expensive, often written in Latin and destined mainly for scholars and university libraries. Few people invested in them. At the same time, however, the book market reacted to the rising levels of literacy. Books in general became less expensive and more geared to the interests of a broader readership. Most people read devotional literature, popular novels and romances, and almanacs containing not only weather forecasts for the year but also miscellaneous prophesies, words of wisdom, popularized tales, and even stories of scientific discoveries.

During the seventeenth century, a popular almanac, often read by entire families, might sell more than eighty thousand copies. Pamphlets were even more commonly available. Flyers and broadsheets (literally: a single sheet of paper) conveyed religious or political messages, usually with large illustrations, in the vernacular, and frequently rhyming. They cost no more than a few pennies and had the added advantages that they could be quickly and clandestinely printed, sold discreetly in the street, easily hidden from prying eyes, and later reused for other purposes. Most of these ephemeral publications have been lost or exist now only in a few copies, and it is impossible to gauge how many copies were produced. We do know, however, that single-sheet publications—which were frequently scurrilous, sacrilegious, partisan, polemical, and often obscene—could be found in different corners of Europe within months of their original publication. They were also extremely popular and much more widely disseminated than books.

Even if we are lacking exact numbers, it is clear that not only the book trade but also the news business entered a new phase after 1600. During the sixteenth century, there were only a few gazettes, usually copied by hand and sent to interested parties. The most important of these, published by the Fugger Bank in Germany, had existed since 1568. By 1620, newspapers appeared regularly in European capitals. They were read mainly by merchants who had a particular interest in knowing about political and other developments elsewhere. They had to make decisions about investments and trade based not on hearsay but on hard information. A trading city like Hamburg even produced two competing publications, and, toward the middle of the century, some of these newspapers had circulations of several thousand copies, which were often passed from hand to hand, multiplying their influence.

Not all contemporaries welcomed this development. The Italian satirist Gregorio Leti wrote:

> Ordinary people . . . read the news as it is written, but interpret it as they wish, and they more often turn good news into bad than bad news into good . . . [while previously] people had no reason to exercise their minds in the delusions and fantasies that they read in the newspapers, but were idle, everyone thinking about his own affairs instead of those of their rulers; but deluding and fantasizing has turned the people into princes, the ignorant into experts, the simpletons into sages, and the obedient into disobedient.[11]

Some rulers eyed education with great suspicion, especially since knowledge tended to be partisan, and educated subjects were more likely to display the disobedience that Leti had criticized. Cardinal

Richelieu in France and King Philip IV of Spain actively engaged in limiting educational possibilities for their subjects, closing schools and colleges that they viewed—not without some justification— as hotbeds of rebellion. Everywhere in Europe, the powerful were engaged in a fruitless struggle to contain the flood of pamphlets and books through censorship and repression, which had the same result as all politics of prohibition: It drove the market underground and made the authors more radical than they had been before, and the publishers, incidentally, richer.

War and political unrest throughout Europe spurred a glut of political publications, propaganda material, and pamphleteering. During the middle of the seventeenth century, some five thousand different *Mazarinades*—pamphlets attacking or ridiculing the powerful Cardinal Mazarin—were printed and circulated in France in editions ranging between a few hundred and thousands of copies. In Catalonia, a political battleground then as now, a dozen or so different pamphlets were appearing every year until about 1620; a generation later, there were more than seventy. In Germany, the Thirty Years' War generated an entire propaganda industry, while the English Civil War (1642–1651) led to an explosion of publishing activity. In 1641 alone, English printers published some two thousand pamphlets as well as three newspapers. The following year saw the publication of twice the number of printed works and more than sixty papers. Observers noticed this change with a mixture of pride and concern, as Edward Chamberlayne wrote in 1657:

> The English since the Reformation are so much given to Literature, that all sorts are generally the most knowing people in the World. They have been so much addicted to Writing, and especially in their own Language, and with so much licence or connivance, that according to the observation of a Learned Man, there have been during our late Troubles and Confusions, *more good* and *more bad Books* printed and published in the English Tongue, than in all the vulgar Languages of Europe.[12]

IDLE TALK AND FABRICATIONS

Let us return to Amsterdam. In the tolerant climate of this premier marketplace for wares and ideas, administered by a succession of pragmatic and sober city councils, the city developed into an international center not only of trade but also of printing and publishing. By the eighteenth century, it would very lucratively supply censored and clandestine books to France and other European countries. But what were people reading in this open-minded city? Toward the end of the sixteenth century, the vast majority of burghers might have possessed a Bible, as expected of good Calvinists, and a handful of religious tracts. In 1578, Mary Gisbertsdochter, for example, the widow of a well-to-do wine tax agent, left behind two books, "one of the legends, the other with the gospels."

Just three decades later, popular novels were all the rage. They were read so widely that a teacher felt it necessary to warn his pupils against "witty and useless books," lest God punish them. In 1612, the bishop of Antwerp even tried to ban beloved romances and comic tales such as *Mariken van Nieumeghen*, *Till Eulenspiegel*, and *Floris and Blancheflour*, while the poet and politician Jacob Cats was worried that his children might become overexcited by reading the story of Doctor Faustus.

These early modern equivalents of "penny dreadfuls" recycled old material: *Till Eulenspiegel*, *Floris and Blancheflour*, *Amadís de Gaula*, and other legends of fools and knights and damsels in distress dated back to the Middle Ages, offering comic, romantic, or heroic stories of reassuring familiarity. But contemporary tales also reached the eager market. The saga of the life and exploits of the Dutch pirate Claes Compaen found an enthusiastic readership, and the foreign tales of the widely traveled captain and merchant Willem IJsbrantszoon Bontekoe sold so well that a new edition was issued every second year.

The demonic Doctor Faustus, meanwhile, had led to a minor horror genre, very much scorned by Calvinist pastors, whose own books had trouble asserting themselves against the frivolous and sensationalist competition. The best the guardians of public morals could do was to lean on the booksellers to include an edifying preface by a religious authority that gave the correct interpretation of these tales of sin—a preface that could be conveniently leafed over by the reader. The many troubled comments recorded by theologians who feared that the simple people were being lured away from a life of piety by immoral tales suggests that they had reason to be concerned. The Dutch had become a reading nation, and books were loved not only by the well-heeled middle classes but also by sailors and maids.

News from around the world was equally popular and easy to come by in a harbor city, and in time of war it was essential, and controversial. Distortions and propaganda in the public sphere are certainly not a twenty-first-century invention. Already, Abbot Wouter Jacobszoon, who had sought refuge in the city in 1572, lamented that pamphlets and news items about the war that could be bought for a few stuivers in the street were nothing but "idle talk and fabrications" to spread lies about the Spanish. He himself decided to take no further notice of the news, because it appeared to be written only "to spread tall tales and fear and loathing in the hearts of good Catholics."[13]

The reading public wanted sensational stories, and a booming market supplied them aplenty. In the mid-seventeenth century, there were four hundred booksellers in Amsterdam, who usually doubled as publishers. Their shops would have been full of bold title pages with ornate illustrations and large print. The binding could be the most expensive part of a book, so books that were available as blocks of paper, without a binding, allowed the readers, many of whom were also collectors, a choice of having their pages wrapped either in cheap hide or even cardboard, or in luxurious, gold-embossed leather. For those without the patience or the knowledge or the pennies to read entire books, there was another market. Paperboys stood on street corners, hawking the latest news sheets and scandalous stories, as one visitor described: "They are running around with almanacs, Ballads and sometimes with crying

'What a miracle! What shocking news!': there are many lazy scoundrels who earn their money like that."[14]

Some of the pamphlets sold on the streets still survive in Amsterdam's city archives. They bear titles such as *A Terrible Murder in Delft!*, *Horrifying Fire in Wilda!*, *Amazing Ghost in Brussels!*, *Come and See, How Three Students in Kloppenburg Violated Two Girls and Killed Four!*, *See How Sixty-Four Wizards Killed More Than a Thousand People and Six Thousand Cattle!* The title page of this newsflash about witchcraft shows two men being hanged with their feet just above a glowing bed of fire.

A WARNING AND A CALL TO REPENT

Fear of witches and wizards made good copy, and the anonymous writers skillfully exploited the uncertainties of an era in which war, winters, and social instability had upset all certainties. Even in peacetime, life could be full of deadly perils. Between 1635 and 1637, Amsterdam was in the grip of a plague epidemic that killed seventeen thousand inhabitants, almost one person in six. Interestingly, this terrible dominion of death coincided with the height of the tulip-mania investment bubble—it would not be the first time that people were willing to risk everything when faced with the uncertainty of their own survival. The previous epidemic, which had killed ten percent of the population, had raged only a decade earlier and had spread widely across Europe. It is estimated that a million Europeans lost their lives from it, the highest number since the Black Death in the fourteenth century.

One of the reasons for the exceptionally high death tolls from these epidemics was the scarcity of food—always after poor harvests, but also generally, due to the constant inflation of grain prices. This particularly affected city dwellers, who had little or no access to land for growing their own vegetables. Many people had little gardens beyond the city walls, but in times of war, these went untended, and they were always subject to the vagaries of the weather. People who were entirely dependent on the market, and on money, to have at least one daily meal of bread or porridge or perhaps a little bone-and-offal broth, were the poorest, living in cramped, unsanitary conditions without access to clean water. They were also the first to die, thus making room for a new wave of migrants from the countryside.

Jean Nicolas de Parival, a scholar at Leiden University who taught French literature and also dealt in French wines, sought an explanation in the classics as he looked in horror at his own era of wars, famines, and epidemics. The Greek historian Hesiod and his Latin admirer Ovid had written about the eternal succession of the ages, from the Golden Age to the Silver and Bronze Ages, and finally to the Age of Iron. This, to Parival, was the outline of his time, as he argued in 1654: "I call it the Age of Iron, because here all Calamities and mysterious events converge, which only appeared occasionally in times past."[15]

Like many of his contemporaries, Parival understood the events of his day as an ominous mixture of divine wrath and a perversion of the natural order, which also, incidentally, indicates that the author knew of telescopic observations, and may even have studied sunspots:

> It is this terrible period about which Scripture writes so clearly, it is the Realm of Iron, which breaks and overwhelms all things. The seven angels have emptied their cornucopias over the earth and they were brimful of blasphemy, misfortune, massacres, injustice, treason and other, infinite calamities. . . . We have seen and we still see how kingdom rises against kingdom, nation against nation, pest, famine, earthquakes, terrible floods and signs in the sun and on the surface of the moon and the stars, fear stalks all countries because of the storms and the roaring of the seas.[16]

In 1602, just one generation earlier, Renward Cysat (1545–1616), the official chronicler of the Swiss town of Luzern, had voiced similar sentiments when describing an earthquake in the Swiss Alps:

The lake makes waves which are towering in the center, the animals are restless. The same inland, in the meadows . . . where people have heard the cattle huddle together and raise the most pitiful noise, they scream and roar against all nature and custom, as if they wanted to inspire humankind to lament its fate.

There are many opinions and guesses abroad about this earth quake, also from scholars and learned professors, and everyone is talking about it, that it is metaphysical and must mean something. The Almighty have pity upon us and help us.[17]

Cysat was an extraordinary man who not only served as local historian but also wrote a book about Japan, and he was one of the first people in Europe to collect local folktales and legends. His curiosity, though, was particularly engaged by the changes he observed in nature. While Parival would later write about a breakdown of civilization and of nature itself, Cysat still trusted in the existence of a divine order:

It might not be thought necessary to describe such things and to expend any effort on them. But while our sins have alas brought upon us [winter] seasons which show themselves to be ever longer and harsher and harder, and we feel a waning of the creatures, be it people, animals, but also fruits and things growing in the earth, also the strange changes in the stars and the air, and we have documented all kinds of memorable things for us and our descendants as a warning and a call to repent.[18]

TEARS TOO PLENTIFUL TO COUNT

Cysat and Parival were separated not only by age but also by generation and historical experience. Both felt and observed the changes in nature, but their views of the world were necessarily influenced by the fact that to Cysat this was a natural catastrophe, whereas thirty years later Parival wrote from a historical horizon dominated by warfare and uprisings, by endless calamities. A great proportion of the Europeans of Central and Northern Europe shared these experiences.

The event framed by these two accounts is the Thirty Years' War (1618–1648)—the bloodiest conflict to have erupted in Europe up to this point—which devastated entire regions of Central Europe and cost the lives of half of their populations. This is why Parival was moved to call it the Age of Iron.

Pious observers still saw God's wrath in the famines, failed harvests, and wars; the latter, if not caused absolutely by climate change, were certainly made worse by it. Hunger, revolts, refugee treks, rape, and plunder became commonplace. Contemporaries understood these events as being linked. The Königsberg professor Simon Dach was one of many poets who penned songs for Protestant worship, songs that often refer to the harsh weather: "Just God, whence will things go / With this times of cold? / What punishment have you prepared / for our perversity? / What will the steady hoar frost and snow / Falling on land and sea / Mean for our poor souls?"[19]

God's just anger caused not only frosts but also the devastation of war, the preachers thundered. The Protestant minister and poet Paul Gerhard (1607–1676), one of Johann Sebastian Bach's favorite lyricists, wrote this hymn especially for times of "great and unseasonal wetness":

> In constant rancour we abide
> And war is ruling far and wide,
> Envy and hatred everywhere,
> In all estates discord and fear.
> That, too, is why the elements,
> Reach out against us with their hands,
> Fear coming from the depth and sea,
> Fear from the very air on high.
> In mourning is the source of joy,
> The sun no longer sends bright rays,
> The clouds are raining like a fount,
> The tears too plentiful to count.[20]

War dominated the lives of a great many Europeans, and it too became an engine of the accelerated transformation of the Continent. War costs money, a great deal of money, and the reduced profits from traditional agriculture were no longer sufficient to finance it. Not only that: Another revolution was occurring. New technologies and tactics not only made warfare considerably more effective but also multiplied the costs of it. Those rulers seeking to dominate Europe or simply to repel an invasion were under tremendous pressure to innovate, not only in terms of weaponry and the training of soldiers but also in finding new sources of revenue.

The key to innovation in weaponry was the improved musket, now with fast-discharging, quick-loaded flintlocks instead of the old and slow wicks, and more accurate artillery, which now cut through the battle lines. The Swabian town of Nördlingen, where Rebecca Lemp and other women had died at the stake a generation earlier, would suffer particularly dramatic effects from these developments. In 1634, this wealthy Protestant community of ten thousand souls, located at a strategically important position in southern Germany, was caught between two fronts of the Thirty Years' War. Five hundred Swedish soldiers were garrisoned there to prevent the troops of the Habsburg Emperor Ferdinand III from capturing the town.

Nördlingen was crucial to the Catholic emperor's campaign to reclaim the region from the Protestants, so Ferdinand sent a force of thirty-three thousand men to lay siege to the town, now heavily fortified. Positioning their artillery on the surrounding hills, the attackers built bastions and trenches and even diverted the local river to cut off their opponents' water supply. The townspeople soon had almost no water to drink, or to put out fires ignited by the hot cannonballs, or to run the water mills needed to grind their grains into flour—all the more critical because the siege force also targeted the horse-pulled mills used to replace them.

The situation in Nördlingen deteriorated very quickly. Great numbers of farmers from surrounding areas had sought refuge within the walls of the town, increasing the pressure on the towns-

Matthäus Merian,
Darstellung der Schlacht bei Nördlingen,
am 6. September, Anno 1634.
(aus: M. Merian, *Theatrum Europaeum*, 1670)

people's resources. Soon the price of flour had risen sharply, leaving the poorest folk exposed to hunger. Then the horses started dying, their rotting bodies producing a terrible stench. Diseases set in. Soon the first human corpses lined the streets—so many, in the end, that there was not enough time, and not enough space, to bury them all before they began to decompose.

The Protestant armies were slow to coordinate. By the time they had converged to do battle with the Catholic besiegers, the latter had had time to prepare thoroughly for the attack. On September 6, 1634, the battle to relieve Nördlingen finally occurred—a brutal and bloody confrontation that proved crucial in the war overall. The battle was illustrated in great detail by the engraver and publisher Matthäus Merian.

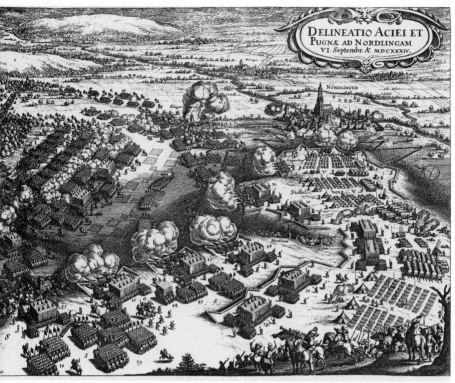

A fateful day: The battle of Nördlingen not only influenced
the course of the Thirty Years' War but also demonstrated
the superiority of new weaponry and tactics.

In the accompanying illustration, the town itself, shown in
the upper right-hand corner, is clearly protected by high medieval
walls, surrounded by a second defensive ring. In front of that, on
the so-called Gallows Hill where Rebecca Lemp had perished in
the flames, is the main encampment of the Catholic army, protected
by earth bastions with fortified advance positions. From here, the
emperor's artillery bombarded not only the town but also the Protestant troops arriving from the left, who were already engaged in
open battle.

The Catholic positions on the plain, in a star shape, are flanked
on both sides by cavalry in dense, square positions, as well as infantry units recognizable by the forests of twenty-foot-long pikes.
Notably, the battle is fought mainly not in close combat but with

fire from artillery and muskets, whose thick smoke hovers above the combatants.

The strategic dominance of the imperial Catholic army resulted in overwhelming victory for Ferdinand's forces. The twenty-five thousand Protestant soldiers on the exposed plain stood no chance against the well-encamped opposition forces. Up to ten thousand Protestant soldiers perished, and four thousand more were taken prisoner, while the besieging Catholic army suffered losses of about thirty-five hundred men.

The victory at Nördlingen was the catalyst for the Catholic reconquest of the entire region, with the Protestant armies decisively weakened and demoralized. The exhausted surviving citizens of the town surrendered. Half of those within the walls had already been killed by hunger, epidemics, or shelling; only about four thousand townspeople were still alive.

THE REVOLUTION OF THE BARREL OF A MUSKET

The battle of Nördlingen was a decisive moment in the Thirty Years' War, and Merian's engraving illustrates how much warfare itself had been transformed by recent technical and strategic innovations.

While European wars during the Renaissance had been won by forces of rugged mercenaries wielding halberds and swords, fighting largely on their own account, victory now was more likely to lie in the hands of engineers, mathematicians, and drill sergeants. Merian had a keen eye for detail and military strategy. Infantry regiments are shown marching in tight squares of pikemen, their long spears towering over the battlefield like forests. Smaller units of musketeers are placed like bastions on the corners, while cavalry units wait on the flanks to attack the enemy infantry. Artillery positions dominate the entire battlefield, ready to bombard the densely packed opposing forces.

This disciplined and highly organized tactic had been developed by Prince Maurice of Orange (1567–1625), one of the most brilliant strategists of his time. (The mathematician and philosopher René Descartes had served under him as an artillery officer.) Prince Maurice had studied in Leiden, where he had become interested in modern weaponry and scientific ideas of warfare. The smaller battle units he developed were more flexible and less vulnerable to artillery fire than long lines of attackers. Importantly, the new muskets had immense advantages over the outdated, cumbersome, and inaccurate old arquebus. They could be aimed easily and fired quickly. In a

Load, aim, fire: Standardized gestures executed on command revolutionized warfare and became an inspiration for industrial processes.

well-trained unit, one row would aim, discharge, and then fall back behind the protecting wall of pikes to reload, while the next row of musketeers would step forward and shoot in a revolving wall of fire.

New and better cannons also played an increasingly important role. If the trajectory of fire was calculated correctly, it could devastate advancing armies even while they were still out of sight, as well as create havoc inside well-defended cities. A few of these expensive guns could turn the tide of battle. In Nördlingen, the imperial troops had had fifty-four pieces of artillery, while the Protestants could marshal only thirty-two, most of which were captured when they were defeated.

The new, modern firearms could be decisive on the battlefield, but they required a completely new approach to fighting. Artillery officers had to be trained mathematicians in order to deploy their firepower accurately. Musketeers could work effectively only if they were well trained and working with standardized equipment, which meant that they had to be drilled and garrisoned, fed, quartered, and remunerated for long periods. War became even more expensive than it had been. Successful armies now required well-drilled soldiers trained in efficient routines learned by rote, sophisticated logistics, smooth supply lines, standardized weaponry, and professional officers with a solid knowledge of tactics to coordinate infantry, artillery, and cavalry into a lethal choreography. The age of industrial warfare had begun.

One of the immediate consequences of this technological revolution was that the size and number of European armies increased dramatically. During the early seventeenth century, the elector Georg Wilhelm of Prussia had commanded an army of nine thousand soldiers. By contrast, his son and heir, Friedrich Wilhelm, could summon a force of eighty thousand men into the field. At the same time, naval forces were expanded, with more and bigger ships, armed with greater numbers of modern guns. Until the Spanish Armada was defeated by Arctic winds in 1588, its twenty-two galleons and 108 refitted merchant vessels were thought to be invincible—certainly for the Dutch, who commanded only twenty-five warships. By 1628, however, the Dutch navy had swelled to 114 larger and better-armed warships; by the middle of the century, its shipyards were producing some two thousand military and civilian vessels every year, and some seventy thousand men were being recruited annually to man them.

Such vast forces were costly enough that military leaders soon became cautious about deploying them. An open battle could bring glory to a general, but the losses might be so great and so expensive that, in the long run, his side might lose the war. Instead of risking men and equipment in ruinous confrontations, strategists now preferred a different approach, as the English administrator Roger Boyle, Earl of Orrery, noted: "Battles do not now decide national quarrels, and expose countries to the pillage of the conquerors as formerly. For we make war more like foxes than lions and you will have twenty sieges for one battle."[21]

Siege warfare became the new style of military conflict, and it, too, was revolutionized. The structure of defensive star-shaped trenches around the besieged town of Nördlingen had been improvised according to this new theory. While the high medieval stone walls of the town were almost impossible to defend against sustained artillery fire, the new defenses were lower, constructed of softer brick and soil capable of absorbing projectiles, and they offered better angles for firing back at besieging troops. This so-called *trace italienne* proved almost impossible to take by storm, allowing the five hundred defenders stationed in the town to hold out at length against thirty thousand attackers.

Nördlingen was exceptional by the standards of the time because the siege lasted only three weeks before an open battle decided the situation. Most sieges dragged on for months or even years. The camp of the besieging army grew into a small, well-organized quasi-town, with a sophisticated system of trenches, bulwarks and outposts, streets, and living quarters. Ten years earlier, in August 1624, the army besieging the Dutch town of Breda had counted forty thousand men, who surrounded the town with two defensive rings, ninety-six fortified encampments, and forty-five artillery positions.

Breda had capitulated after a nine-month siege, during which fifteen thousand citizens had lost their lives.

A victorious campaign, or even the capacity to deter potential invaders, required not only vast armies and fleets but also experts trained to control the complex logistics, constructions, tactics, supplies, and accounts, as well as vast amounts of treasure. Perhaps war really is the father of all things. During the seventeenth century, the old feudal order proved unable to cope with the demands on their royal and ducal coffers, and rulers across Europe were obliged to find new ways of financing their campaigns. "A crisis of income of the ruling class ends the middle ages and opens the modern period," commented the French historian Marc Bloch.[22]

Alongside the threat of war, the threat of harsh weather continued. "A third of the world has perished," noted the abbess of the famed Port-Royal-des-Champs convent in Paris in 1654.[23] The Habsburg lands, France, Flanders and the Low Countries, Germany, Russia, as well as parts of Scandinavia and Italy, all endured a seemingly endless series of wars and rebellions, with hunger and exhaustion leaving their populations vulnerable to waves of epidemics.

The crisis of the Little Ice Age had been carving its imprint on European lives for more than half a century. But a subtle and important change now marked the beginning of a new era. While two generations earlier, the response to failing harvests had been religious, expressing itself in the form of public processions, fire-and-brimstone sermons, witch trials, and fantastical invocations of the Last Judgment and the End of Days, an increasing number of intellectuals had begun to think in different, more modern ways about the problems of their time. Some of them, arguing in purely economic terms, tried to find new sources of revenue for their rulers. Others cast about for philosophical answers to questions that could no longer be understood in theological terms. Both of these groups—economists ("moral and political philosophers") and metaphysicians and logicians along with them—were about to catapult Europe into a new kind of social and intellectual reality.

SELL MORE TO STRANGERS

In a world in which political interests were asserted by war, every state was forced to project military power, whether or not it was interested in territorial gains. Most rulers had to enter into alliances and to field considerable armies, even if only for purposes of defense. Financing war became the main concern of government.

For the first time in Western history, scholars and administrators began to think methodically about the structures and possibilities of their society and its economy without relying on biblical injunctions, theological arguments by the doctors of the Church, or even the comparatively liberating philosophy of classical antiquity. Instead, they began to form their theories out of perceived current earthly needs—money for the soldiers, for instance—and on the immediate material givens: geographical, demographic, and economic realities. They were leaving the Middle Ages behind and preparing the ground for what would eventually be called the Enlightenment.

Traditionally, rulers in need of money had simply raised taxes. But with the disruption of agricultural production owing to climate change, and the consequent inflation of grain prices, added to the ravages of frequent war, the scope for more taxation was limited. Any significant increase risked public resistance and even outright rebellion. Already, taxes in France and Eastern Europe had been doubled, forcing countless farmers to simply give up and move into the cities. It quickly became clear that new and different sources of income had to be found.

The solution to this dilemma was simple, argued Thomas Mun, director of the London East India Company: "The ordinary means therefore to encrease our wealth and treasure is by Forraign Trade, wherein wee must ever observe this rule; to sell more to strangers yearly than wee consume of theirs in value."[24] In other words, only a trade surplus could guarantee that taxes and tariffs would enrich the state.

Having directed the powerful East India Company since 1615, Mun (1571–1641) held a key position in the English economy. Inter-

national trade was booming, always in competition with the explod-
ing naval and economic power of the Dutch, the Englishmen's main
rival. Heavily armed and often escorted by warships, trading vessels
warily eyed every sail on the horizon—buccaneers and pirates as
well as hostile powers made international commerce very much a
continuation of war by other means.

Mun was aware of these dangers, and of the potential to his
own country of economic power farther afield. His influential work
England's Treasure by Forraign Trade was published in 1664, more than
thirty years after Mun had written it as an internal political memo-
randum. In these pages, he argued for a new kind of political and
military success, based on aggressive trade policies. He also painted
a society in which economic considerations were to trump all others. The strength of a state, he wrote, was expressed through its
ability to win wars. A powerful prince could find new glory through
the tireless efforts of his merchants:

> . . . a mighty Prince whose dominions are great and united, his
> Subjects many and Loyal, his Countries rich both by nature
> and traffique, his Victuals and warlike provisions plentiful and
> ready, his situation easy to offend others, and difficult to be
> invaded, his harbors good, his Navy strong, his alliance pow-
> erfull, and his ordinary revenues sufficient, royally to support
> the Majesty of his State, besides a reasonable sum which may
> be advanc'd to lay up yearly in treasure for future occasions:
> shall not all these blessings (being well ordered) enable a Prince
> against the suddain invasion of any mighty enemy, without
> imposing those extraordinary and heavy taxes? shall not the
> wealthy and loyal subjects of such a great and just Prince main-
> tain his Honour and their own Liberties with life and goods,
> alwayes supplying the Treasure of their Soveraign, untill by a
> well ordered War he may inforce a happy Peace?[25]

This happy peace was to be built on international commerce and
local employment, most importantly through building up a strong

navy, impregnable fortresses, grain stores sufficient to cover one year of drought, banks to facilitate investment, the training of "Collonels, Captains, Souldiers, Commanders, Mariners," and stores for ammunition and military supplies, "all which will make them to be feared abroad, and loved at home, especially if care be taken that all (as neer as possible) be made out of the Matter and Manufacture of their own subjects . . . for a Prince (in this case) is like the stomach in the body, which if it cease to digest and distribute to the other members, it doth no sooner corrupt them, but it destroyes it self."[26]

The elite as the stomach of society—Mun may have taken this image from Shakespeare, who had used it in his drama *Coriolanus*, where the Roman aristocrats are trying to calm the commoners who are rioting against high bread prices. In many ways, this was an early version of "trickle-down" economics: "Again, the pomp of Buildings, Apparel, and the like, in the Nobility, Gentry, and other able persons, cannot impoverish the Kingdome; if it be done with curious and costly works upon our Materials, and by our own people, it will maintain the poor with the purse of the rich, which is the best distribution of the Common-wealth."[27]

This blessed vision, however, was endangered by the serious character flaws exhibited by the monarch's subjects, especially the poorer ones, who appeared sadly unable to cope with even the most modest trappings of wealth. While the enemy across the Channel was pious and fearsomely disciplined, the English appeared, to Mun at least, increasingly lazy and debauched:

> . . . whilest we leave our wonted honourable exercises and studies, following our pleasures, and of late years besotting our selves with pipe and pot, in a beastly manner, sucking smoak, and drinking healths, until death stares many in the face; the said Dutch have well-neer left this swinish vice, and taken up our wonted valour, which we have often so well performed both by Sea and Land, and particularly in their defence, although they are not now so thankful as to acknowledge the same. The summ of all is this, that the general leprosie of our Piping, Potting, Feast-

ing, Fashions, and mis-spending of our time in Idleness and Pleasure (contary to the Law of God, and the use of other Nations)
hath made us effeminate in our bodies, weak in our knowledg,
poor in our Treasure, declined in our Valour, unfortunate in our
Enterprises, and contemned by our Enemies.[28]

The wealthy merchant Mun clearly saw wealth as a danger to weak
characters, and his therapy was equally unambiguous: "As plenty
and power doe make a nation vicious and improvident, so pen-
ury and want doe make a people wise and industrious." Wealth,
it seemed, was good only in the hands of a small number of people who were born into it, or whose exceptional personal qualities
enabled them to use it well.

A virtuously poor and industrious population ruled by an elite
celebrating "bounty and pomp" appeared to Mun as the ideal order
of society. Only healthy consumption could fill the royal coffers,
which now depended not only on agricultural productivity but also
on trade and industry. Exporting wares and importing gold while
energizing the domestic market was not only the best, but the
only recipe for creating a happy kingdom. Economic thinkers and
administrators in France, Italy, Spain, Germany, and Russia agreed
with him, even if they were frequently less able to realize this vision.
Export on a grand scale without corresponding import of goods
was bound to create conflicts of interest. What one monarch earned
was lost to another, trade was seen as a zero-sum game. Like the
battlefield, the market knew only winners and losers.

With its heavy emphasis on the economy, Mun's idea of society
has some very modern aspects. Karl Polanyi's fear that Western societies could one day shrivel to a mere appendage of the market has its
origins here. At the same time, Polanyi wrote eloquently about the
changes that an economy built on trade surpluses and strong markets will produce within the societies harboring them.

In order to achieve the security and predictability necessary for
investments and contracts, clear laws and strong institutions have
to be created. Courts of law will hand down judgments from which

not even the king and the nobility are exempt; life is increasingly governed not by religious but by man-made rules. The outlines of the modern state are born with the establishment of courts and market regulatory bodies; with the Creation and maintenance of roads, harbors, bridges, and other means of reliable transport and communication; and with a government seeking to attract investment by emphasizing stability.

In a social order based less on feudal hierarchies and more on markets and trade, the emerging middle classes and what would become the working class had important and different roles to play. Members of the middle class were significant not only as consumers of domestic production but also as educated experts who enabled the state to run smoothly and find new avenues of creating wealth. Coming from a position of little assured privilege, their self-interest would support a market that was independent of control from the monarch, and institutions that were capable of protecting them from the whims of noblemen. This created social and economic attitudes that later would be associated with the moral philosopher and economist Adam Smith and with the Enlightenment. Stability and growth of the national economy would increasingly be seen as criteria for measuring the success of societies. This success would not be measured in honor or divine grace, as in medieval times, but in tax revenue, infrastructure, battleships, and munitions depots brimming with equipment.

The poor, too, had a crucial role to play, as Thomas Mun explained. Their cheap labor would transform domestic raw material into expensive articles for export. By paying them little, their employers contributed to higher overall tax revenue, a patriotic act in itself. This also had a moral dimension. Mun and others argued that the poor were undisciplined and thus liable to abandon themselves to fornication, alcohol, card games, and laziness. Only strict work and a frugal life could keep them on the path of morality and godliness; left to their own devices, they would plunge the entire country into an abyss of lawlessness and debauchery.

Mun and his fellow mercantilists vociferously fought to control poor people themselves, but not against poverty. A population on the breadline represented a vast reservoir of cheap workers. Just as in the medieval world, Mun felt, poverty had a purpose: It was an integral part of a functioning society. While the middle classes were supposed to move upward and gain some independent wealth, those at the bottom of society had to be controlled and, if necessary, disciplined. They should not be kept so hungry that they could no longer work effectively, but if they were given too much money, or educated to develop ideas above their lowly station, the whole edifice of society would come crashing down.

THE STATE AS MACHINE

In France, senior administrators and scholars also adhered to and implemented mercantilist ideas—people such as legal scholar Jean Bodin, economist Barthélemy de Laffemas, the dramatist and economist Antoine de Montchrestien, and Maximilien de Béthune, Duke of Sully (1560–1641), the hugely influential finance minister to King Henri IV.

The life of Maximilien de Béthune exemplifies several important currents of the time. As the son of an aristocratic Protestant family, he was able to attend one of the country's premier schools in Paris, but since his family was relatively poor, he had to put his education to use to earn a living. He joined the army, distinguished himself in battle, and became a trusted adviser of Prince Henri of Navarre, the future king.

During his time as an artillery officer, Béthune also served under the Prince of Orange, then Maurice of Nassau, whose innovative military tactics impressed the young Frenchman. This new, scientific, effective style of warfare relied on rational planning, a kind of industrial logic. Béthune thought it could also have applications in other contexts. He returned to France, and, after an advantageous marriage, he soon rose to a position of great wealth and power as an officer, negotiator, royal adviser, and, not incidentally, war profiteer.

Béthune was a pragmatist. In spite of his own deeply held Protestant beliefs, he advised his friend Henri of Navarre to convert to Catholicism in order to receive sufficient support from powerful Catholic aristocrats and thus unify the war-torn country. "Paris is worth a Mass," sighed Henri, and acquiesced. In 1598, the newly crowned King Henri IV of France named his faithful friend Superintendent of Finances. One year later, Béthune became Grand Master of Artillery, Grand Voyeur de France (which translates, slightly disappointingly, as Chief Street Builder of France), and Superintendent of Fortifications. In addition to the nation's purse strings, almost the entire military and civilian infrastructure of the country was now under his control.

As France's top administrator, Béthune was convinced that the country had to be able to project power, and that this power could be achieved only through efficient organization, an extensive investment program, and a strong economy. On his order, modern fortresses were built along the borders, as well as a network of roads and canals. The army was used to assert royal power over remote and lawless regions of the country, and a campaign against highwaymen secured the roads for commerce. But the finance minister thought even further ahead. He reduced the levies on farmers, campaigned for the adoption of new and more efficient agricultural techniques, and drained swamplands to increase agricultural production. He also literally sowed the seeds of a strong economic and military future by having elm trees planted alongside the newly constructed roads. Their trunks would one day become masts on the ships of a powerful French navy.

In his reforming zeal, Béthune uncompromisingly curtailed old tax privileges, abolished sinecures and internal tolls, and did everything possible to lower unnecessary costs, making bitter enemies along the way. The king's trust, however, made him untouchable—or almost untouchable, for he once barely managed to escape an assassination attempt. In 1606, the monarch designated his friend Duke of Sully in recognition of the fact that the state finances had been turned around, and that the formerly deeply indebted Crown was now making sizable profits every year, despite lowered taxes. New industries were contributing to this bounty, especially the silk industry, for which thousands of mulberry trees had been planted to feed millions of silkworms. Four years later, in 1610, Henri IV was assassinated. Deprived of royal protection, Sully retreated from public life.

Mun and Sully were only two of many scholars, merchants, and men of state who supported mercantilist ideas and put them into practice. The first and highest goal had to be a positive balance in foreign trade. In order to boost exports, imports were saddled with special taxes and punitive tolls throughout the Continent, and everywhere high productivity coupled with low wages required a large population competing for this work. Having been depopulated in large part by the Thirty Years' War, Prussia pursued a very active immigration policy. Skilled immigrants were lured with special tax privileges, and thousands of Huguenot Protestants from France made this choice, carrying with them not only their Bibles but also their expertise in silk production and other industries, as well as knowledge of the emerging sciences.

If the revolution away from agrarian subsistence economies and toward market-centered trading and state-directed manufacturing for export proved successful for the government, it was not necessarily so for the landless poor, whose plight had in many cases declined more sharply than ever. The consolidation of large estates producing for the market, and the enclosures of the commons, had made countless families homeless. Towns and cities swallowed desperate new arrivals, first in slums and later in paupers' graves. An army of poor men in search of work were turned into tiny cogs of the great economic machine, working in primitive manufacturing and in the navy and the army.

Of course, this large-scale exploitation in the name of economic growth affected not only the poor in Europe itself. To feed the rising consumption of an emerging middle class at home, seafaring nations, such as England, France, Spain, the Netherlands, and Portugal, seized raw materials and exploited slave labor in the colonies and dependent territories they created around the world. Simultaneously, local manufacture in the exploited territories was undermined in order to provide easy markets for products from home.

In conjunction with slavery—at this point mainly practiced on the Caribbean plantations—the system of long-distance exploitation became a crucial link in the chain of wealth creation that turned Europe into the most powerful region of the world. Together with military subjugation and economic oppression, it also introduced European modes of government as well as Christianity, spread by missionaries with tremendous zeal and frequently with ruthless violence.

A PROFITABLE TRADE

Trade in the subjugated territories followed the logic of political power of the time. Trade was viewed as a form of warfare, so that it seemed morally acceptable to exchange comparatively cheap goods such as tools or small weapons, or even worthless trinkets, for precious metals, expensive spices, or other valuable luxury goods: silk, porcelain, exotic woods, indigo, and tea.

If the producers of these goods were slaves, the balance sheets in Europe made even better reading. The cost of purchasing and feeding a slave was minimal compared with his or her profitability over a working lifespan of five to ten years, for few slaves survived longer. During this time, most were subjected to a merciless regime of hard physical labor, while being fed and housed and clothed in the poorest way and often punished with extreme cruelty for real or imagined infringements of the rules imposed upon them.

This was the beginning of the highly lucrative "triangular trade," which would expand steadily over the course of the next two centuries. Cheap goods and arms were sent to the west coast of Africa to pay African slavers for cargoes of Africans captured in the interior and herded to the coast for transport across the Atlantic. The sea crossing—"the middle passage"—itself was brutal, to such

Every sack stained by human blood: Slaves in the West Indies
were producing cane sugar for the European market.

an extent that it was calculated as a business loss that one in every
five out of this living cargo of densely packed bodies below decks
would not survive it. Hunger, thirst, epidemics, and violence would
further reduce a slave's chance of surviving, even before the ship
reached harbor.

Slave traders in Bordeaux and Nantes, London and Plymouth,
Amsterdam and Antwerp, Porto and Naples found that they could
easily write off such losses. Their ships would come back packed
with slave-produced tobacco, coffee, sugar, cotton, and other luxury goods. Business was booming. The new city-dwelling middle
layer of society was hungry for these luxuries and willing to pay
extravagant sums for them. They set the fashion and added to the
chance for social distinction. Better-off European consumers were
themselves also quite willing to invest spare sums in the companies
engaged in the slave trade.

How many African, South American, and Asian slaves were

laboring on plantations, in mines, and in factories during the seventeenth century? Not as many as in subsequent centuries, but the slave trade, even at this point, involved tens of thousands and then hundreds of thousands of forced workers, although their precise numbers cannot be known and their fates would not all have been the same. Especially in Asia, it was at times difficult to determine who was a slave, an indentured laborer, a serf, a bonded worker, or even a captured prisoner. European traders were generally careful with their financial records, but most of these have since been lost. The documents that survive, however, paint a clear picture of the rising importance of slavery and other violent colonial practices for European businesses, for state taxes, and for the wider economy as a whole. A few examples may give a sense of how significant the slave trade had become.

In 1619, the first enslaved Africans were herded off the deck of a Dutch merchant ship in Jamestown, Virginia, and sold to the owner of a tobacco plantation. In 1700, the English colonies in North America alone had some twenty-eight thousand slaves living in them. In the Caribbean islands, the introduction of sugarcane around the middle of the seventeenth century had created a huge demand for cheap labor. In 1650, an African slave cost roughly six pounds on the Caribbean slave market. With a life expectancy of five to ten years, until he would die of malnutrition, overwork, or sickness, the slave could harvest great quantities of sugarcane that could be sold back to Europe at immense profit. At that time, some twenty percent of the inhabitants of the West Indies were slaves; fifty years later, around 1700, this number had risen to eighty percent. Thousands of enslaved Africans were also working at this time in Southern Europe, particularly in Portugal, Sicily, and Naples, where almost twenty percent of the population had African origins.

Europe's huge economic success was often presented as a victorious campaign against the enemies of true Christian virtues— whether rival trading nations, foreign rulers, indigenous populations, or transported slaves. The truth, of course, was different. The spectacular career of one Dutch merchant, Jan Pieterszoon Coen (1587–1629), shows just how a trading empire could be established.

A man of burning ambition, Coen was only twenty when he boarded ship as a junior merchant of the powerful VOC (Vereenigde Oostindische Compagnie), the Dutch East India Company. His deeply held Calvinist beliefs ring through every sentence of his detailed reports back to headquarters in Amsterdam, where his shrewd suggestions and ruthless execution of company policy soon drew the attention of his bosses.

Coen constantly lobbied to establish a more muscular presence in Southeast Asia in order to fend off claims from rival powers such as Portugal and England, and to construct a large military base in Jakarta, then ruled by a sultan who seemed amenable to a deal with the Dutch. "In India, we cannot in my opinion prevail without authority. . . . We must assert ourselves with force of arms, or we can whistle through our empty fingers," noted Coen.[29]

On Coen's second trip to Indonesia in 1612, the newly minted senior merchant began carrying out his plans. The directors in Amsterdam were so delighted with his progress that they promoted him to a director's position. After inconclusive negotiations with the sultan, however, Coen grew frustrated. He firmly believed in the mercantilist idea of trade as war by other means, and his Indonesian opponents were refusing to bend to his demands. In 1619, he ordered an army of roughly a thousand men to attack the undefended city of Jakarta and burn it to the ground. Hundreds of residents were murdered during the ensuing massacres, but there was no lamenting on the part of the VOC. On the smoking ruins of old Jakarta, Jan Coen, the newly appointed governor-general of the province, built the Batavia fortress, a feat that would not have been possible, as he acknowledged to his fellow directors, "had not the Almighty fought on our side and blessed us."[30]

Two years later, Coen again found it necessary to make good the Lord's blessings with cool steel. He himself led a penal expedition to the Banda Islands, famous for their high-quality nutmeg. Local traders had broken the terms the VOC had imposed on them by

God with us: In the name of the Lord and of profit,
Jan Pieterszoon Coen spread fear and death.

repeatedly selling nutmeg to English and Portuguese merchants. In retaliation, and to set an example, Coen had all the nutmeg merchants publicly executed. Then he drove the population into the mountainous interior of the islands, where they were hunted until they starved. Some fifteen thousand villagers died to underscore Dutch trading might—and those who survived, including the women and children, were sold on local slave markets.

It was Jan Coen who successfully established the Dutch claim to Indonesia, thus accumulating vast profits for his company and for himself. He was the first in a long line of pitiless governors, all ready to despoil the country and its population in the pursuit of profits for their own kind. We have some idea of the numbers of people executed, murdered, starved to death, and enslaved, but the full human cost of this regime—and so many like it—will never be known.

THE CURSE OF SILVER

Mercantilist theorists demanded that the royal coffers be filled with gold and silver in order to finance military expansion on the part of the state itself or for defense against land-hungry neighbors. Most silver arrived in Europe via Spain, which controlled rich mines in Peru. But even the influx of wealth from the colonies, and the rise of professional administrations at home, proved insufficient to settle the finances of many European states, with the Duke of Sully in France one of the few to be able to turn the national economy of his country into a profitable concern. We owe much to the decades-long research by French historian Fernand Braudel for our picture of this seventeenth-century world that may seem familiar to us today, where the rich grew richer and the poor fell deeper into poverty, and where the prices of essential goods rose, especially in the cities. Despite all the Peruvian silver, and partly because of it, inflation had become a problem.

The mechanism of this increase in the cost of living was simple: As profits from agriculture had become less certain and grain prices had increased with every failed harvest, governments increasingly required payment of taxes, tariffs, and tolls not in kind—as had once been common—but in money. This, however, also meant that governments lost control of grain prices, a vital lever in maintaining social peace. Grain prices were now more likely to be determined by speculative investments on Europe's stock exchanges in London, Amsterdam, and Venice. Grain was a commodity that could be traded across borders and priced according to competing demands. Agriculture had become part of a market in which goods were bought and sold—not as local circumstances required but in order to make profits for people often very far away. For local populations, this market development generally resulted in more goods for those wealthy enough to participate, and in more frequent hunger for those too poor to take part.

The effects of this systemic change toward market economies and a monetarized, increasingly speculative economic system, could be manifold—nowhere more dramatically than in the case of Habsburg Spain, the most powerful and the richest country in the world at the end of the sixteenth century.

The downfall of this vast Catholic empire, spread across the world, was perversely caused in part by the excessive wealth created by the immense influx of Peruvian silver. Habsburg Spain was a static, traditionalist, not to say hidebound society, in which the Church played a supreme role and aristocrats were largely preoccupied with maintaining their hierarchical privileges and observing the customary rituals of their estates. Questions of precedence at international events, or even at formal dinners at home, might be discussed for weeks in advance.

Regional rulers remained preoccupied with asserting their power over their own domains. Spain was crisscrossed by local customs borders, and special levies were raised on all kinds of goods and produce, so that it was impossible to create a national trading network. In addition to this, the silver flooding into the country was beginning to distort prices. The wool producers of Castile, for example, were unable to sell their wares to the weavers of Catalonia without incurring punitive tariffs, nor could their wares be exported, since they were too expensive to be competitive internationally. The development of an urban middle class, which was fermenting change in other European countries, was made almost impossible in Spain by restrictions on economic activity, by the political power of the aristocracy, and by the Catholic Church. The Church's retrogressive policy toward education resulted in the closure of schools and universities deemed seedbeds of mental and spiritual corruption—at the same time that so many new places of learning were being established elsewhere in Europe.

In 1609, to cap off this strict regime, King Philip III decided to expel the Moriscos—merchants who had converted to Christianity from Islam—in whose hands much of the Mediterranean trading networks with Muslim countries had flourished for centuries. The Moriscos were accused of collaborating with pirates, secretly practicing Islamic rituals, and sympathizing with Spain's Muslim enemy, the Turkish Ottoman Empire. The virulent Archbishop of Valencia dubbed the Moriscos godless heretics, claiming that he was looking forward to the day when he could confiscate all their possessions and use the men themselves as galley slaves in the king's Catholic navy.

The expulsion of the Moriscos apparently was not very popular with the Spanish people at large. Italian mercenaries had to be hired to deport the Morisco families. Within three years, 300,000 people had been deported, most of them to North Africa, where they had never lived. A personal catastrophe for those deported, it was also a disaster for the Spanish economy. One-third of the population of Valencia and a fifth of the inhabitants of Aragon had been Moriscos, as well as tens of thousands of people in Andalusia. Losing them meant that Spain lost a vital class of merchants and many crucial trading connections.

For a while, the silver coming in from the colonies was enough to compensate for growing structural imbalances. Peruvian mines financed several wars for Spain, as well as palaces and luxury goods for its elite. But the silver was undermining the value of the local currency, and leading to high inflation; perversely, as the silver flowed in, the royal treasury fell more heavily into debt.

In 1607, with the silver shipments still arriving from across the oceans, the royal house was declared bankrupt, for the fourth time in fifty years. To add insult to financial injury, in 1628 half the Spanish treasure fleet was captured by the combined efforts of Dutch admiral Piet Pieterszoon Hein and the Sephardic-Jewish pirate Moses Cohen Henriques. The pair made a fortune for themselves, with the Dutch government itself claiming more than eleven million guilders, enough to pay its army for eight months.

But it was above all Spain's cultural and intellectual rigidity that drove its relentless decline. As the mightiest nation on earth, it had seemingly had no need to adapt to changing economic and social circumstances, but within less than a century, the former superpower was relegated to Europe's margins, where it would remain for some four hundred years.

Those who controlled the economies of other European countries had put their faith in mercantilist ideas. To keep their treasuries filled, they strove for economic growth based on the systematic exploitation of the natural world and of the poor and vulnerable at home and overseas. Several institutions arose or were transformed in the wake of this mercantilism: boards of trade, law courts, legal frameworks, orderly state powers, and the professions related to them. This rising class of traders and experts had their own, often independent views of the world; they were ready to see their own self-interest placed on a new philosophical foundation.

Philosophical ideas began to reflect the interests of the new middle class, and in their turn they would rationalize and justify the societal changes occurring because of it—philosophy and society in effect shaking hands. Time and again, we see intense connections linking modern warfare, mercantilism, internationalized markets, and a rising middle class, with groundbreaking ideas about knowledge, religion, science, nature, the state, and the rights of the individual.

OFFICER, RETIRED

In Sweden, by the middle of the seventeenth century, winters were on average two degrees Celsius colder than the twentieth-century average. Climate reconstructions show that seven of the ten coldest years of the previous five centuries fell between 1569 (seven degrees below average) and 1614 (five degrees). It seems that one of these murderous winters may have cost the continent one of its most brilliant thinkers.

In the autumn of 1649, a private scholar and former artillery officer was taking leave of his friends in the Dutch town of Sandpoort. Tears welled in his eyes as he boarded the coach to set off for Stockholm, a distant place where he had accepted a prestigious new position. René Descartes (1596–1650) was one of the new men; his intellectual adventurousness was regarded with hostile distrust by most churchmen and many princes, too, because they threatened to upset the supposedly divinely ordained social order. Descartes had spent the previous twenty years in exile from his native France in the more liberal Netherlands; now he was hoping for a significant improvement in his lot. As tutor to the brilliant young Queen Christina of Sweden, he would be the undisputed intellectual star at the court of the Protestant monarch, a secure position from which he would be able to further his philosophical research.

Descartes was an ambitious man. The third child of a lawyer to the provincial lower aristocracy, he had profited from the recent educational revolution by becoming a pupil at the Jesuit college of La Flèche in northern France. Like his father, he had then studied law, though without much conviction. Looking for a life more exciting than that of a provincial lawyer, the young man traveled to the Netherlands and joined the army of Maurice of Nassau, later the Prince of Orange.

For a young graduate from a good family, the army offered the chance for a solid career and a quick rise through the ranks, with considerable economic opportunities in addition to military fame. But another motivation may have been even greater in Descartes's mind: As an artillery officer, he could indulge his passion for mathematics. Building fortresses and trenches, and especially calculating the trajectory of artillery fire, required a good knowledge of geometry. Those with such skills could quickly become indispensable to their superior officers, who typically had come from the higher nobility and had purchased their army commissions, without having had any systematic military training.

Under Maurice of Nassau, the young Descartes encountered the scientific, rational approach to warfare, in which calculation, logistics, reconnaissance, and espionage were infinitely more important than physical courage in battle or athleticism on horseback. Here, Descartes, a rather timid man whose immense daring was purely intellectual, was in his element. He even had enough time on his hands to pursue his interests in the natural sciences. In the Dutch town of Breda, he encountered the scholar Isaac Beeckman, and there he wrote his first work, *Musicae compendium* (1618); it reveals a young author in search of an intellectual challenge, and a rigorous scientific method.

After his military experience in the Netherlands, Descartes changed sides and joined the Catholic imperial army of Maximilian I of Bavaria, which brought him into contact with the work of two preeminent scientific minds of the day. In Regensburg, he visited the observatory of Johannes Kepler, and it is even possible that he met the great astronomer himself. Moving with the army, Descartes then had a chance to visit the workplace of the Dane Tycho Brahe, another founding father of modern astronomy. During the perishingly cold winter of 1619–20, when the entire imperial force was able to travel across the frozen Danube River, Descartes remained holed up in a small room with a roaring fire. Already in November, the frosts had been so biting that he had retreated to his heated room—his "stove," as he called it—and stayed inside for three days. But his need to retreat was not only due to the weather. Extraordinary new ideas were taking shape in his mind. According to his early biographers, he had three visions here, in which a divine messenger revealed to him the contours of a new philosophy.

These visions may be nothing but a biographer's flourish ascribing hard, intellectual work to celestial grace. Whatever they were, their appearance changed Descartes's life. Now twenty-four years old, he ended his military career and in the spring of 1620 decided to dedicate himself entirely to "natural philosophy"—that is, to science. He sealed his decision with a conventional yet telling gesture: He made a pilgrimage to the Italian town of Loreto.

Over the following years, Descartes traveled through Europe, visiting scientists, and searching for his own intellectual vocation. He ended this peripatetic life in 1625 and moved to Paris, where his growing reputation as a debater and outstanding mathematician opened professional and social doors for him. His dedication, though, did not prevent him from leading a full and even somewhat dangerous life: It is documented that he won at least one duel during this time.

Descartes's intellectual ambition propelled him onward. The intellectual life of Paris at this time was carried on not at the famous Sorbonne University but in a handful of aristocratic salons, where guests were able to speak freely, without having to fear police spies or censorship. New ideas, still largely for the leisured elite behind closed doors, could not safely be discussed in public. Science took place in private studies and laboratories. At the Sorbonne, students could choose only among theology, law, and ancient languages, and all printed works were subject to strict censorship by a commission of priests.

The Netherlands offered far greater intellectual freedom, and not only in the universities, so Descartes decided to move there. The remote northerly region of Friesland was the first stop for the philosopher in search of a new home. Soon he moved to Leiden, where he refined his mathematical skills and indulged his interest in astronomy at the university. From there, he went on to Amsterdam, where, among other things, he fathered a daughter with his housemaid, Helena Jans van der Strom. The little girl, Francine, died at the age of just five, leaving the father distraught, as several letters testify. The tragic loss of his daughter would be a second turning point in Descartes's life and work. While his earlier interests had been rather general, and focused on mathematics and astronomy, after 1640 he dedicated himself more singlemindedly to the fundamental questions of philosophy.

His first major work appeared in 1641. *Meditations on the First Phi-*

losophy investigates questions he had already discussed four years earlier in his *Discourse on the Method*, but it delves more deeply into matters of meaning and knowledge. In both of these texts, however, Descartes reveals his Promethean ambition.

He was preoccupied with the foundations of knowledge, the possibility of knowing that something is true and accurate, and he was attempting nothing less than to find one unassailable, rational principle on which all human knowledge could be grounded. Knowledge comes from the senses, Aristotle had argued, but what if we are all just dreaming a world that exists only in our dreams? How can we be certain that instead of a benevolent God, there is a malicious demon making us believe that the illusion in which we are living is objective reality? Is there a true and indisputable first principle from which real and objective knowledge about the world can be constructed?

Descartes famously found this Archimedean point in the simple sentence *Cogito, ergo sum*—I think, therefore I am. In order to reach beyond illusion, we can and must doubt everything to divest ourselves of seductive beliefs, he argued—all perceptions and memories, everything we believe to know and to be true, everything that can be doubted. But at the same time, we cannot deny that this very act proves the existence of a Thinking Thing aware of its own doubt. The thinking, self-aware mind is proof of its own existence; radical doubt creates unassailable certainty. This is the rock, he explained, on which all knowledge can be built.

Rooting truth in awareness immediately confronted Descartes with a problem. While there is no denying the existence of a thinking mind reflecting on itself and on the impressions and memories brought to it by perception, how can this mind separate true perceptions from false ones? And how can it be certain that there is indeed a world outside itself, a world that can be accurately represented by the thinking mind? If I know that I exist because I think, how can I be certain there is anything in the world besides my thinking self? The specter of solipsism arises, the ultimate loneliness. As a famous philosophical example puts it: If a man awakes from the dream in which he was a butterfly, how can he be certain that he is a man who dreamed about being a butterfly, and not a butterfly dreaming that he is a man?

For Descartes, everything—the possibility of all philosophy— hinged on this question. If he wanted to establish true, rational knowledge, he had to discover or construct a solid bridge between the thinking mind and an objective world outside it, about which the mind could gain accurate information. He had to prove that there was a world that existed independently of thought or perception, and that this world could be known and understood adequately by human intelligence. He had to prove that we do not owe our percep- tions to the malicious caprice of an evil demiurge, but rather that the world is indeed what we perceive it to be.

Descartes had set out to doubt everything and to assume noth- ing by using a strictly rational, even mathematical method of build- ing an argument from first principles, but to take the next, crucial step, the Jesuit-trained philosopher fell back on theology. Indeed, there is a surprisingly large number of theological echoes in his work. Even his *Cogito, ergo sum* harks back to Saint Augustine, the great intellectual of the early Church: How is it possible to know that we are receiving true information about the world, and that the image we construct in our mind is accurate? For centuries, philoso- phers had struggled to find an answer to this foundational question. With supreme confidence, Descartes introduced them to what he understood as the answer.

His proof, which has the undeniable elegance of a mathemat- ical argument, takes only a few sentences. Our mind can imagine all kinds of things. Most of what we imagine consists of dreams, illusions, falsehoods, chimeras, monsters, and demons of our emo- tional existence; all seem to exist clearly in our mind's eye. Among all these ideas—most of which have no correlation in the outside world—one stands out because it is simple and necessary: the idea of a Perfect Being, which is dictated not by our fertile imagination but by logic itself. A Perfect Being must really exist, because if it had no existence, it would be imperfect. Therefore, a Perfect Being exists, unlimited in time and space (otherwise, its perfection would be destroyed), and hence omnipresent and eternal, the origin of all being, the creator of this temporal, material world. Moreover, the idea of perfection also means that a Perfect Being must be truthful, because any lie would be falsehood, incoherence, imperfection, and therefore irreconcilable with its perfect nature.

From this graceful swerve, Descartes deduced that a divine creator not only exists necessarily, but that he has created the world out of his truth and thus, because all deceit is foreign to his being, he reveals the world to his creatures as it really is. The lonely *Cogito* finds a home in the real world, in the rock-solid certainty that it can reach true knowledge about the world. This argument, Descartes believed, had ended one of humankind's oldest and most difficult debates in one fell swoop—by the simple application of scientific reasoning. Better than that: Almost in passing, he had also proved the existence of an omniscient, omnipotent, and benevolent God from reason alone.

Again there is a theological echo here, and it leads to a tantalizing question. The proof of God's existence that Descartes used was known to scholastic philosophers four hundred years earlier. The argument that God's existence is necessary because the idea of his perfection implies his existence—the "ontological proof" to theologians—had enjoyed a considerable career since it had been propagated during the thirteenth century by Anselm of Canterbury. But even some of Anselm's contemporaries had refuted this proof, with perhaps the nicest of these refutations coming from the monk Gaunilo of Marmoutiers, who cheekily asked whether, if he could conceive of a perfect island in the ocean, this island would also necessarily have to exist. With this simple question, he had drawn a sharp distinction between "perfection" and "existence" as attributes of an idea, and the existence of facts outside of language. Something can be necessary, he implied, because logic and grammar may dictate it, but that need have no bearing on what is out there in the world, outside of language.

Having been schooled by Jesuits and being widely read in the theological and philosophical literature of his day, Descartes surely knew that the linchpin of his proof of the possibility of objective knowledge of the world outside ourselves had been removed generations earlier. His crowning argument did not even hang by a silk thread—it was left dangling in midair. Even though theologians were unlikely to point out this obvious weakness, it is quite impossible that Descartes believed he could fool his highly educated readership throughout Europe. So why did he allow this to stand? Was it simply because he could conceive of no other argument capable of holding his system together?

There is at least one other, and more intriguing, possibility. Seventeenth-century authors, even in the Netherlands, had to be careful about what to commit to print. Ideas that were deemed heretical by the authorities would at best spell the end of a promising career in the civil service or the Church, but they might also force the writer into exile, or even cost him his life. Working in many countries under more or less draconian censorship, philosophers and other writers often chose to work with rhetorical ruses and strategies, hints to sophisticated readers, allowing them to grasp the author's true intentions and arguments while leaving simpler minds convinced of the surface argument. Radical and controversial ideas were very often projected between the lines of a text, rather than in the words as they literally appeared.

For scholars, it was normal to expect books to be written with a powerful and craftily wrought subtext. Next to the usual allusions and quotations (often designed to make a curious reader consulting the original text stumble across the lines preceding or following the quoted ones), one favorite device was the elaborate and orthodox refutation of a heretical idea. The refutation would be posited in a manner so obviously weak and incoherent, and would give the original so much space to shine, that many readers would find themselves agreeing with the dangerous idea that a seemingly incompetent author had attacked. Was Descartes playing a similar game here? Did he want his readers to draw the obvious conclusion that if this argument does not hold, then it is impossible to prove the existence of God, that there can be no certain knowledge about the world, and that in the end all experience may be a mere dream?

That would be the very opposite of what Descartes claimed to say, and there is no indication in his correspondence that he entertained atheist ideas, but it is also true that he was a cautious man who, after learning of Galileo's trial in Rome, decided not to publish his masterwork during his lifetime. The hidden motivations and contradictions of seventeenth-century intellectual life leave its results still shimmering with ambiguity.

Ambition and caution ruled the philosopher's life, and they almost certainly motivated the long northward journey he began in September 1649. His ideas were beginning to fall foul of the Dutch Calvinist authorities, and he thought it might be wise to leave the Netherlands for a while. Sweden's young and impetuous Queen Christina, who was burning to meet him, offered him a post at court as her personal tutor. The invitation provided a seemingly ideal way out.

After weeks in drafty, rattling coaches, Descartes finally reached Stockholm and met the astonishing monarch, who enjoyed a Europe-wide reputation as an intellectual prodigy. Soon, however, the aging philosopher was experiencing the drawbacks of serving this energetic and eccentric queen. Christina demanded that he write a libretto for a ballet, but mostly she appeared to ignore him, temporarily obsessed as she was with the study of ancient Greek. The urbane and well-traveled philosopher quickly became bored: "It seems to me that people's thoughts freeze here in winter, just as the water does," he confided in a letter to a friend in warmer climes.[31]

Soon, however, the sparkling young queen turned her attention to the philosopher, and Descartes found that it was more attention than he could absorb. She demanded that he give her private lessons in her unheated library every morning at five o'clock—Christina

Philosophy on a frosty morning: Descartes and Queen Christina of Sweden.

was not susceptible to the cold, and required no fire to be set there. It was winter by now, and Descartes was a delicate man who preferred to remain in his bed in his well-heated room until after midday. During these very early morning audiences, he was obliged to attend the queen standing, with his head bared to the winter air. This regime quickly proved too much for him. On February 11, 1650, he died of pneumonia in Stockholm.

※

PERHAPS DESCARTES'S HEROIC attempt to put all human knowledge and all thought on a compelling, logical basis was doomed to fail from the outset—like every attempt before and since. Still, it marks an intellectual sea change: the confident claim that reason, empirical knowledge, and logic alone can get to the heart of things, that human beings are inherently rational, and their makeup and place in the world can therefore be fully explained. During the early days of the scientific revolution sweeping through Europe, this position was the main starting point for elaborating a process through which the world could be grasped and described according to a method agreed on by all participants and subject to empirical verification. Science, as we understand it today, was making its appearance. Whereas Bible passages had always furnished the ultimate proof of truth or falsehood, from now on, theories about the material world would be tested by experience—and experiment.

It would be a long time, however, before this method was widely agreed upon. For the time being, there were different kinds of knowledge competing to replace the fading authority of faith and scripture. Never had Europe's intellectual landscape been richer in philosophers and necromancers, Rosicrucians and mathematicians, occultists and alchemists, magicians and soothsayers, engineers and astrologers, mediums, Kabbalists, and prophets. Their laboratories and libraries became the testing grounds of new ways of understanding the world, mixing mathematics and myth, stargazing and classical literature, empirical observation and stagecraft, to form a whole cosmos of ideas and intellectual experiments. Dramatic climatic change had disrupted the natural order so sharply that the need for new explanations was now immense. Prophets always thrive in times of upheaval and uncertainty. Many of those who tried to provide new answers were simply charlatans, but from among those who were sincere emerged the framework of the intellectual world we still inhabit today.

THE SUBVERSIVE REPUBLIC OF LETTERS

Knowledge needs networks, and the frequently unexpected connections among very different seventeenth-century thinkers illustrate both the fluidity of their debates and the dangers of engaging
in them. Rational and empirical thinking, the beginnings of a
scientific method, were often violently opposed by churchmen
alarmed at the suggestion that the key to all truth was slipping
from their grasp. If the intensified interest in the observation of
nature spurred new avenues of thought, it also galvanized powerful enemies of change.

Marin Mersenne (1588–1648), the century's supreme net-
worker, was a crucial link in many of the debates of his day. Like
his friend Descartes, Mersenne was a graduate of the Jesuit school
at La Flèche, but unlike most thinkers of his time, he had been born
in rural poverty and had worked his way into an intellectual life.
Besides his own scientific research in mathematics and acoustics,
and his extensive writings on theology, he became, in himself, a hub
of European intellectual life. Not only did he correspond with hundreds of scientists and thinkers, he also copied and disseminated
their works, discussed the works of others, and created connections
among people who had been toiling in relative isolation.

Mediating between contradictions that also marked his own life,
Mersenne was a perfect embodiment of a time swept up in intellectual transformation and uncertain of the shape of things to come.
He was a priest and a theologian who battled Kabbalists, Rosicrucians, and other mystical thinkers with great determination, as well
as a promoter of empirical knowledge and logical deduction. A
pious man in constant correspondence with some of the greatest
and most fearless scientific minds of his time, he did his utmost to
disseminate their ideas. Among his friends were philosophers such
as his former schoolmate Descartes, the English political theorist
Thomas Hobbes, and the materialist Pierre Gassendi; Mersenne
corresponded with the mathematician Fermat and with leading scientific thinkers such as Galileo and the robustly empirical Constan-

CHA
RI
TAS

Marin Mersenne
de L'ordre des peres Minimes

The internet of the seventeenth century: Marin Mersenne
corresponded with numerous scholars and spread new ideas.

tijn Huygens. He tutored the brilliant mathematician and theologian
Blaise Pascal, and he traveled to Italy no fewer than fifteen times in
search of new books and new ideas. But none of this prevented him
from welcoming the cruel execution of Lucilio Vanini, a notorious
heretic, in 1619 in Toulouse.

We can chart the outer limits of Mersenne's intellectual tolerance by looking at Vanini's short and tragic career on the edge of
what could be said (and printed). He, too, grappled with the contradictions between faith and rational inquiry. Born in 1585 in

Lecce, Italy, into the family of a wealthy merchant, Vanini studied in Naples and became a Carmelite monk, calling himself Fra Gabriele. Financially independent after the death of his father, the young friar pursued his studies in Padua, where he became involved in the dangerous world of local politics. After choosing the losing side, he fled to England, living incognito in the palace of the Archbishop of Canterbury, head of the Anglican Church, who had extended his protection to the brilliant young Italian.

In 1612, a grateful Vanini converted to Anglicanism, but he was soon troubled by new theological doubts, and finally he appealed to the pope himself to gain readmission to the Church. When the pope granted his request, the archbishop became furious and tried to have his guest arrested. Having traveled to England as a refugee from political persecution, Vanini now found himself in danger of being brought to trial there as a heretic. The Gunpowder Plot, in which a group of Catholic conspirators had tried to kill the king and blow up Parliament, had been uncovered only eight years earlier, and anti-Catholic sentiment was running high. As a traitor to Anglicanism and a personal enemy of the archbishop, he knew that he could face the gallows. Only another plot, this time involving the Venetian ambassador, allowed Vanini to flee from prison. A refugee once more, Vanini arrived in Paris in 1614. He was twenty-nine.

Vanini had gained readmission to the Catholic Church, but he had also gained a good deal of notoriety. His career as a scholar seemed finished, all chances of employment blocked, his very life in danger. In an effort to regain the trust of the Church hierarchy, the young scholar traveled to Rome to justify himself before the pope, but when one of his travel companions was arrested by the Inquisition in Genoa, Vanini fled back to France. The following year, he published a work against atheism and skepticism in order to strengthen his case, and this strategy appears to have worked: His next book appeared with the official approval of the doctors of the Sorbonne, the guardians of theological orthodoxy.

Unfortunately, the learned gentlemen did not appear to have read the volume they had endorsed so roundly, a work with the inconspicuous title *De Admirandis Naturae Reginae Deaeque Mortalium Arcanis* (On the Wonderful Secrets of Nature, Queen and Goddess of the Mortals). Behind the ostensible praise of God's Creation lay another, more dangerous argument, a skeptical tone, and an array of troubling ideas that Vanini had disclaimed and denied in his text with suspicious irony. More shocking still: The fictional, edifying dialogues between the author and a young man named Alessandro could be taken to suggest that the entire universe is purely material and that there is no such thing as an immortal soul.

Vanini was a virtuoso at dancing with his arguments, bringing them to the fore and walking them back, dismissing them and rediscovering them later, asking unexpected questions. No one before him had devoted an entire philosophical dialogue to the question of tickling. Time and again, he offered materialist, empirical explanations for natural phenomena, only to cover them with a thin and brittle theological varnish. The book's two dialogue partners roam through the entire landscape of knowledge about the world. They discuss comets and lightning, snow and rain, the energy of an arrow shot from a crossbow, the origins of mountains and islands, earthquakes and magnetism, sperm and reproduction, sensory organs and illusions, heathen images and miraculous healings, witchcraft and dreams.

By asking the kinds of questions a five-year-old might ask, Vanini circumnavigates orthodox responses and exposed weaknesses in the theological account of the world. The author's alter ego, Julius, develops another view in which even the sky consists of matter and the earth is round—a vast ball moving according to immutable natural laws, not through the hand of God or His angels. "Had I not been raised by Christians in school, I would claim that the sky is moved like an animal, moved by its own form, which is a soul. The mass of the skies is moved, like the elements, in a circle, by its own form."[32]

A godless universe moving according to its own laws, circling around nothingness? Alessandro cannot believe his ears. When he demands further explanations from his teacher, Julius adds: "Everything finite can perish. The sky is finite, therefore it will perish."

Time and again, Vanini employs the time-honored stratagem of seemingly knocking down materialist or atheist arguments, but making such a mess of it that his opponent has the more convincing case. Asked about the origin of animals, Julius mentions the then-fashionable theory that small creatures gestate spontaneously in the mud; then he turns his attention to the origins of humanity:

> Others dreamed, that the first human came forth out of sluggish apes, pigs and frogs, because human behaviour is similar to that of these animals. But some less radical atheists say, that only the Ethiopian is the issue of apes, because a special warmth can be found in both.[33]

Julius pursues this thought, ostensibly in order to be able to refute it more completely. "The atheists pretend that the first humans walked on all fours like animals," he scoffs, but he remains fascinated by heredity. His own father, he writes, was already old when he sired his son and communicated his semen "only with moderate powers and very slowly." Therefore, the author concludes:

> I was born with little power or strength. If I have a generous way of thinking, a pleasant figure and a body with few weaknesses this is because my father, even though old, was still lively and happy at times and because a young woman warmed his old and frosty limbs.[34]

According to Julius, everything from the universe as a whole to the genesis of an individual can be explained through nature itself. Virtue and vice? Nothing but the mutual effect of "semen and the imagination," the influence of the stars, and the ingredients of one's food. Good eating leads to a virtuous life:

> Feeling is a tool of reason and every acting thing functions according to the nature of the tool. That is why among the lowest people, e.g. sailors, coachmen, barrel makers and day labourers,

etc., one can find the most godless people. They are wild, inhospitable, fearless, have no respect for religion, because they eat rough and bad fare, which creates thick blood and thick, restless minds.

Having exposed himself to accusations of materialism, the saving swerve follows immediately, even though it is formulated with a suspicious dose of irony: "But I want to leave these follies aside. I do not want to counsel these godless people to buy themselves pontic roots and other medicines which take away the melancholic and bile-like juices, but I enjoin them to be assiduous in using the Christian sacraments, which the Apostolic Church administers justly."[35]

In the last of the fifty dialogues between Julius and Alessandro, Vanini seems almost unable to control his satirical bent, or his enjoyment in setting traps for his readers. God as ultimate goal of all existence? "Nonsense!" exclaims Julius, because then humans would be more important than their god. "Then—is everything meaningless?" asks the student fearfully. "The heavens protect us from Epicureanism," the teacher erupts, before retreating to the official position. "But what about all the sufferings in the world? Are they not sent to try us?" asks Alessandro, reiterating his belief in the immortality of the soul. "Bravo," Julius comments drily. "All animals have a desire for permanence, they want to live on through their offspring and be remembered by them."[36]

Julius, however, goes even further in his musings about religion. In classical antiquity, he says, the wisest men believed that the only religion worth following was the one without dogma, given to us by nature itself.

The other laws, they said, were nothing but confabulations and deception introduced not by the devil . . . but by princes to lead their subjects more easily, and by priests, to win money and honour; not sealed by miracles, but through copied writings whose original cannot be found anywhere. By this means . . . that is, through fear of the highest, omniscient being which pays back everything with eternal rewards and punishments, the common plebs (*rustica plebecula*) is held obedient and subservient.[37]

It comes as little surprise that the doctors of the Sorbonne soon became aware of their error. Vanini's book was condemned by the Inquisition and handed over to the hangman to be publicly burned on the Place de Grève. Once more, the author had to vanish quietly, and quickly. Unable to return to Italy or England, he hid in a monastery in Brittany, where he worked on a radical reinvention of himself.

Several months later, a mysterious Italian created a stir in southern France. Named Pompeo or Pomponio Uciglio, he enjoyed the protection of Duke Henri II de Montmorency, the king's godson, and a man known as much for his unorthodox views as for his equally unorthodox private life. Even the free-spirited duke, however, soon found his protégé untenable. Uciglio was pursued by rumors, and he had an apparently insatiable hunger for life. He was a *libertin*, a man without morals, said his detractors; he had nothing but disdain for moral decency and participated in scandalous, unbridled orgies. Not everyone, however, was repelled by this description, and the young Italian soon found another aristocratic employer, the Comte de Caraman, a man also known for his conflicted relationship with conventional morality.

Just as things were beginning to improve for the ex-monk living under a false name, something decisive must have happened, even though his true identity was not yet known by those around him. Little detail is known, but Vanini suddenly was arrested by the Inquisition. It appears that his behavior had enraged a straight-laced councilor in Toulouse. The annals of the Toulouse city hall document what followed:

On Thursday, the second day of the month of August . . . in the house of the heirs of the late Monhalles, in the capitoulat of Daurade, the gentlemen Olivier and Virazel made a prisoner and took him to the city hall. The young man said that he was thirty-four

years of age, born in Naples, in Italy. He calls himself Pomponio
Ucigli and he stands accused of having taught atheism.[38]

This time, there was no escape. Still unrecognized, Vanini was
handed over to a worldly court and found guilty of atheism, immorality against nature, and seducing the young. He was sentenced to
be taken to the Place du Salin in Toulouse, where his tongue was
to be ripped out, after which he was to be strangled and burned.
The sentence was carried out on February 9, 1618. Witnesses of the
execution said that they were unable to forget the dying man's ago-
nized screams.

><

MARIN MERSENNE HAD never met Vanini, but he knew him by reputation. Now he showed himself delighted with the execution of a
godless man and lost soul who had shared so many ideas with him:
"The libertins are spreading in France," he wrote. "They have not
spent enough time thinking about the example of Giordano Bruno,
and they have made Vanini their leader and proclaimed his new
position."[39]

The contradictions in Mersenne's attitude become clear in light
of his friendship with another priest and scientist, Pierre Gassendi
(1592–1655), a man of the cloth but also perhaps the most important materialist thinker of his generation. In his parish in the little
town of Digne, at the foot of the Alps, the parishioners knew their
priest as a man who was stupendously learned and eminently able as
well as seemingly pious, but one who had never had a career worthy
of his considerable gifts. He was an eccentric, constantly embroiled
in correspondences with other scholars, getting involved in all manner of arguments and intellectual disputes, perpetually bombarded
with scientific papers, and preoccupied with his own observations
and research.

The camouflage of the provincial eccentric may have been useful
to Gassendi, whose intellectual brilliance and commitment made
him a key figure of his time. His life had been characterized by curi-

PETRVS GASSENDVS DINIENSIS z.
Hic est Ille, dedit cui se Natura videndam,
Et Sophia æternas cui reserauit opes:
Inuida non totum rapuistis Sidera Vultum
Naniolias, Mentem pagina docta refert.

A world made of atoms: the priest Pierre Gassendi thought more
highly of logic and evidence than of the dogma of his church.

osity and restlessness. As a young man, he had traveled to Paris and
then to Flanders, where he met the scientist Isaac Beeckman (who
had earlier welcomed Descartes). He also met, and admired, the
exiled Thomas Hobbes.

Digne was a quiet town, too quiet for the restless priest, who regularly went to Paris to meet his friend Mersenne and other scientists, visit scientific meetings and academies, or accept invitations to salons, where he socialized with such famous *libertins érudits* as the poet Cyrano de Bergerac and the young Molière. Here Gassendi was part of a world very different from his native place and from the day-to-day worries of his parishioners. Behind closed doors, everything could be said, and was said. Mersenne, who estimated that there were fifty thousand atheists living in Paris at the time, would have been deeply distressed to know his friend was spending time in such wicked company. Gassendi, though, relished debate and radical dissent. The *libertins* were rationalists, skeptics, hedonists, Epicureans, materialists—with a few deists thrown in to keep the conversation interesting.

Most of these people were too prudent to publish their thoughts—Vanini's gruesome fate was still fresh in their minds. Gassendi, though, found ways to make his thoughts as clear as possible without risking his life. Supplied by Mersenne with scientific works, letters, and ideas, he became an authoritative voice on all matters theoretical, from physics to philosophy. All across Europe, his views on current debates and publications were feared and eagerly anticipated.

When Descartes published his *Meditations on First Philosophy*, it was Gassendi who led the discussion among the scholars who were debating the book's central claim—namely, to put all human knowledge on a rational and irrefutable basis and thus establish an objective and verifiable standard of truth. Gassendi analyzed the argument of his famous colleague and carefully ripped it to shreds. He himself had invested a good deal of time into thinking about what can really be known, and whether we can ever reach a true account of the world. About this, and about Descartes's argument, he remained resolutely skeptical.

Gassendi deconstructed the *Meditations* on every level. He demonstrated that Descartes's "proof" of God's existence was not only logically flawed, it also could not be valid, because all knowledge about the world must originate in our sensory perception, and nothing that comes from any other source can be called knowledge. This strikingly simple idea has frightening consequences: If all knowledge comes from the senses, then what if they are indeed unreliable or even mendacious witnesses? That is possible, Gassendi conceded, but repeated observation by different observers can still establish a high degree of certainty. It may be impossible to determine whether two people who are burned by an open flame experience their pain in exactly the same way, but their experience and the evidence of it—burned flesh—are enough evidence for avoiding open flames.

But if all knowledge comes from the senses, what about knowing things that do not reveal themselves to our senses? Impossible, judged the priest, taking sides in this debate. Nothing can be known without sensory experience, not even God. Especially not God. Descartes's attempt to prove, through logic, the existence of a being beyond sensory perception, crumbled under the beam of Gassendi's analysis. You can believe whatever you like, the priest implied, but you can only know something if you can or could experience it through your physical perception.

These ideas were influenced by the skeptics and Epicureans of classical antiquity. Gassendi was fond of quoting them and obviously considered himself in the tradition of authors such as Lucretius. The world, the Frenchman argued time and again, consists of atoms, of tiny parts of matter that can connect and react to one another in myriad different ways. There is nothing outside the material world, or almost nothing, the priest noted almost guiltily, as he prepared a somewhat hastily put-together appendix to his philosophy of matter—an appendix large enough to contain the heavens.

There was an obvious contradiction between the assumption of a material universe about which nothing can be known that cannot be perceived, and the Catholic vision of a heaven populated with angels, an army of bribable saints, and a God who could interfere in the course of history. But Gassendi managed to find a way to connect these two aspects of his thinking, or at least to allow them to coexist. The bridge between these two worlds was shaky, but the

author could not break with the Church, and frequently he emphasized the divine inspiration of the Bible and the importance of the
sacraments and obedience—while at the same time defending his
learned friend Galileo, after his condemnation by the Inquisition.

Like Galileo, Gassendi was an enthusiastic and capable astronomer, who observed and recorded solar eclipses and other notable
phenomena, probably with a telescope built with lenses sent to him
by Galileo himself. Still, the priest observed certain boundaries in
his scientific works. On the one hand, he rejected the heliocentric
model of Copernicus because it could not be reconciled with biblical revelation; on the other, he calculated the trajectories of the
planets and repeated Galileo's experiments on gravity.

While Gassendi never fully succeeded in reconciling the dogma
of his Church with his observations and philosophical convictions,
he also fought, together with his pious friend Mersenne, against all
forms of mysticism. In a debate with the English doctor and occultist Robert Fludd (1574–1637), Gassendi led a double-pronged attack,
using arguments from science and from theology.

To put it mildly, Fludd's thinking was eclectic—an early example
of the New Age approach of picking the tastiest morsels from very
different spiritual and mystical traditions and baking them into a
single spiritual pie. There was a little Kabbalah culled from the only
available translation by Johannes Reuchlin, some mysteries of the
Rosicrucians, a smattering of Paracelsus. These ingredients formed
a mystical vision in which a world spirit, creator of light and of the
Earth, was living inside the sun. Fludd treated his London patients
not only with pills and tinctures but also with prayers and incantations, an approach that might well have been as effective as any
other at the time.

Gassendi attacked the idea that other religious or spiritual traditions were on a par with Christianity, denied that Fludd was com-
petent as a theologian, and also argued that only empirical proof
could create actual knowledge. In spite of this strong assault,
though, Fludd's views retained considerable influence among thinkers of later generations. The idea that different cultures and tradi-

tions could be investigated not from the perspective of someone who was already in possession of the ultimate truth, but also could be regarded as equally interesting and equally true or false, only became more fascinating as the debate about truth and revelation unfolded. Montaigne would have been pleased.

Pierre Gassendi was only one of the scholars attempting to find peaceful coexistence between revelation and empiricism; others included the astronomer Johannes Kepler, who wrote not only on the hexagonal structure of snow crystals but also on theological questions; Isaac Newton, who spent his later years almost exclusively trying to uncover the mysteries of scripture by engaging in numerical mysticism; and the astonishing German Jesuit Athanasius Kircher in Rome, who was fully involved not only in researching geological and archaeological questions but also in trying to justify his work by quoting from the Gospels.

GERMANUS INCREDIBILIS

Athanasius Kircher's life has an impressively epic quality. Born in 1601 or 1602 as the ninth child of a simple farming family, he was sent to the Jesuit college in Fulda, Germany, where he also studied Hebrew with a rabbi. In 1622, his theological studies were interrupted by the horror of the Thirty Years' War, and he was forced to flee. En route, he almost drowned as he tried to cross the frozen Rhine on foot and broke through the ice. Eventually, he reached a safe haven in Germany; here he would stay for another ten years before accepting a position at the Vatican, a post that seems to have been created especially to allow his vast intellect to flourish.

Kircher's mental empire resembles an immense kaleidoscope in which diverse elements recombine in different constellations, creating new figures and new meaning. He mastered almost all the areas of knowledge his era had to offer: He worked on deciphering Egyptian hieroglyphs and attempted to construct a perpetuum mobile; he researched volcanoes and abseiled into the crater of Mount Vesuvius

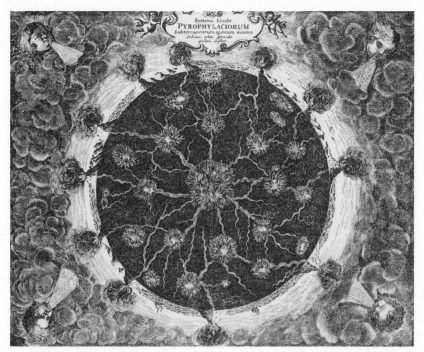

A look into the interior of the earth: No mystery was too great for the
intellectually adventurous Athanasius Kircher.

to study the forces below the surface of the Earth; he taught mathematics, physics, and Oriental languages; he wrote about music theory and studied droplets of blood under a microscope to understand how epidemics spread; he searched for the lost continent of Atlantis and speculated that Adam and Eve had spoken ancient Egyptian; he wrote an important work about Chinese culture (the Jesuits maintained a strong presence in Beijing); and he estimated the size of Noah's Ark according to its need to contain all known creatures and food and water for forty days. In his *Mundus Subterraneus*, he suggested a new understanding of the tides, posited that the interior of the Earth was filled with raging fires, and attempted to reconcile the existence of fossils with the biblical account of the Creation.

No problem was too great or too intimidating, no idea too trivial for Kircher's insatiable interest. He developed a new and somewhat sadistic musical instrument, a cat organ, in which cats would be arranged in a row according to the pitch of their meows, and the player could pinch their tails to make them "sing." The instrument was never built, but in his Museum Kircherianum at the Vatican, amazed visitors could inspect other marvels, including his famous collection of Egyptian and Greek works of art, stuffed and dried animals and plants from the New World, and an outstanding library that seemingly contained all worldly and spiritual knowledge.

Kircher's brilliant mind was legendary, but shortly after his death his work began to be regarded as riddled with errors and baseless speculation, the work of an eccentric intoxicated by the boundless possibilities of knowledge. His ideas about hieroglyphs were revealed to be as misguided as his beliefs about subterranean oceans and corridors of fire running through the Earth. His empiricism, which he had always emphasized, broke down under the weight of his fantastical theories. His syncretistic ideas about ancient Egypt and the true identities of Adam and Eve, Moses, and the "Egyptian Oedipus" worried the Church hierarchy and were mercilessly mocked by other scholars. Pierre Gassendi, incidentally, was also among his many correspondents, but he never engaged his fellow priest in public debate.

During these momentous years of the mid-seventeenth century, the members of the self-described Republic of Letters were struggling to define the cornerstones of scientific thought. Whereas earlier generations had looked to Aristotle, Plato, or Saint Augustine as ultimate arbiters of truth, the power of empirical observation was slowly beginning to replace the earlier models. This process took time, and had to contend with the formidable opposition of what Diderot would later call "the conspiracy of magistrates and priests." Even scientific thinkers such as Gassendi still found it necessary to appeal to a higher, religious truth—and many of them did so with full conviction. The radical doubter Descartes, after all, had followed his famous *Cogito, ergo sum* with a theological argument.

It was not only arguments and standards of truth, however, that were undergoing a deep transformation. Knowledge itself was changing in structure, in extent, in kind, in method, and in ambition. For many contemporaries, this process proved irresistible. The contradictions between traditional, religious, and dogmatic truths and the truths of the new scientific world were glaring, and many fine minds failed in their attempts to reconcile them. In many cases, the appeal and the simple effectiveness of evidence-based deductive thinking proved to be stronger in the long run, but at the beginning of this process, everything seemed possible.

This debate brought many citizens of the Republic of Letters to the edge of the metaphysical abyss. Gassendi insisted on the truth and central importance of the Gospels, but an attentive reader could not escape his contradictory argument: that all knowledge must come through the senses, even if they prove deceptive, and that nothing at all can sensibly be said about any reality inaccessible to the senses. That means: Nothing is certain, we have no access to a transcendental truth, we are alone. A priest could write about the true faith as long as he pleased; his philosophical arguments spoke a different and unmistakable language.

The guardians of the old faith had been alarmed for some time by this new tone in the intellectual debate, but now they saw that the debate was taking place among people who previously would have been excluded from it. The leading figures of this radical conversation did not live in monasteries or at court. They hailed from humble farms (Mersenne, Kircher, and Gassendi), from the houses of merchants (Vanini, Spinoza), or were trained in the law (Descartes) or other middle-class professions. Knowledge was passing into the hands of Europe's social inferiors. Edward Hyde, an English courtier who chose to disregard his own bourgeois background when he became first Earl of Clarendon, bewailed this terrible development in his *History of the Rebellion and Civil Wars in England*: "Dirty people of no name" had had the effrontery to formulate their own views about the world based on nothing but the strength of their logic.

These "dirty people of no name," members of a rising middle of society, were actors of a stunning and frequently frightening degree of energy. Even those beliefs and institutions hallowed by age or regarded as sacred were no longer immune from being doubted, challenged, and even discarded, though some members of this "troublesome class" also felt the emotional loss of old certainties. But they were quintessentially urban people, and life in the city followed different laws from those existing in the countryside.

Great metropolises such as London, Paris, Amsterdam, Naples, Vienna, and Prague attracted countless migrants into their expanding slums. Refugees, fortune hunters, itinerant workers, established burghers, and passing visitors lived cheek by jowl, and with them their languages, religions, convictions, and customs. Ports and market cities in particular offered different perspectives and new experiments in living, as well as information about distant continents and cultures. There were new luxury goods, new books, pamphlets, newspapers and broadsheets, a fashion carousel, and sophisticated patterns of consumption and novelty for the more fortunate members of society, as well as jobs in a bewildering new variety of trades and professions.

This city world had made consumers out of its denizens; whereas country people produced for themselves almost everything they needed, city folk did not. Food and shelter, clothes and entertainments—all were purchased for money. City dwellers were entirely dependent on the money they earned and therefore also more vulnerable to the inflation caused by rising grain prices. Survival was possible only for those who earned enough. Those who were not able to earn at least a few pennies every day were thrown on the mercy of institutionalized charities, herded into workhouses, or left to starve by the roadside. The city offered much, but it dealt mercilessly with those unable to keep up with its pace.

VIRTUE IN THE DROWNING CELL

Life in Amsterdam offers a perfect example of this pitiless dynamic, partly because its approach to the world seemed to be justified by its breathtaking success as the city expanded so rapidly, bursting with people and money, with self-confidence, possibility, innovation, and new wealth.

In a dizzying two generations, Amsterdam had grown from a provincial backwater to one of the centers of global trade and also of intellectual daring. Having laid the foundation of its stunning wealth by trading in Baltic grain, the merchant patricians of Amsterdam had soon conquered huge territories overseas in a campaign of militarized trading. This strategy proved successful beyond their wildest dreams, and it demonstrated the power of the new model of economic growth based on exploitation.

Trade with the Baltic seaports was particularly successful because the grain had been cultivated by serfs who received little or no payment, much like the indigenous plantation workers, slaves, and prisoners tending the fields and working in factories in Indonesia, Guyana, and Surinam, producing pepper, nutmeg, indigo, tea, coffee, tobacco, and other luxury items. The work of these unpaid or barely paid people underwrote palaces and cathedrals, parliaments and courthouses, theaters and universities, and armed forces, including those employed to force them to keep working.

Without apparent irony, Amsterdam's Protestant city fathers insisted that only an industrious life could be pleasing in God's eyes. Far from admiring the nobility in other, court-centered countries, the burghers despised aristocrats for their unproductive laziness. This attitude toward work as a good in itself was quite new in Europe, where not having to work had long been seen as a sign of wealth, or even nobility. In the new Protestant merchant cities, however, there was no place for idlers.

For those of the poor who were considered unacceptably idle, Amsterdam had its own special institution: the Rasphuis (literally: sawing house), a place symptomatic of the mentality of the growing city and its thrusting self-regard. The Rasphuis was a special kind of prison: half reform school, where indolent young men were taught the value and necessity of hard work, and half repository for lazy individuals too old to be reeducated. All inmates had to work to earn their keep. In a special workshop, they ground tropical woods to dust that would be used for dyeing fabrics. They were paid, but idling and other infractions were punished by a range of sanctions

No one shall eat without working: The Rasphuis was Amsterdam's very
Protestant answer to indigence and poverty.

that included canings in front of a paying public, townspeople who
looked on to satisfy their curiosity, and no doubt, too, to reinforce
their own determination to keep doing an honest day's work.

Several visitors to the Rasphuis reported on a special punitive
feature for which no archaeological record exists: a cell in the base-
ment used for any prisoner who absolutely refused to work. The cell
would slowly fill up with water, and the prisoner was equipped with
a pump. His choice was simple: work or drown.

Industriousness is next to godliness, the Protestant theologian
Calvin had taught, and he had found eager disciples among the
pragmatic and hardworking Dutch traders.

Calvin's doctrine of predestination strongly influenced the way
the poor were valued and treated in the Protestant world. If a sinner
could force God to forgive him simply by being repentant and liv-

ing morally, it was thought, this would imply that the Lord was not omnipotent, because he would be coercible. God saves a soul, said Calvin, because it is his pleasure to do so, not because of something that person does or does not do; and since God is omniscient, it follows from this that God has chosen whether or not a person will be saved even before that person is born. God shows this divine "predestination" by granting wealth to those he favors with his grace. Therefore, said Calvin, being rich implied having being chosen, being especially beloved by God. If you made money, it showed that God loved you. But if you remained poor, you had to admit that you had not been chosen.

This theological argument tallied conveniently with the self-image of the merchant patricians whose proud houses lined the canals of Amsterdam's city center. Ultimately, it meant that their wealth was willed by the Creator himself, and their profits were further proof of his blessings. In Catholic France, the dramatist Molière made a career out of skewering the pretensions of these pre-destined nouveaux riches, but in the Protestant North, the connection between God's grace and growing material wealth fit perfectly.

It is interesting to notice the similarities and the differences between this argument (very compelling if you happened to be rich) and the justification used by the aristocracy to assert their own claim to power. In medieval societies, the static, pyramidal structure of society was held to be God-given, and the position of the nobility was seen as part of this divine order. Monarchs ruled by God's grace, and ultimately everyone had been allotted a position by God's infinite wisdom, hence each position was largely unchangeable: Once a king or a peasant, always a king or a peasant. Calvin and his Catholic counterpart Cornelius Jansen, however, had offered a theological justification for social climbing. God had chosen them to become rich and powerful, regardless of where they had been born.

In most of Europe, the newly rich imitated the taste and style of the aristocracy. The Netherlands, once again, was the exception; here, a bourgeois aesthetic began to assert itself as early as the

seventeenth century. Successful Dutch merchants and professionals did not want to hold any kind of court, even if they could have
afforded to do so, and they did not need countless retainers and a
large entourage. As far as they were concerned, their wealth was the
product of hard work, and they were proud of it; they showed the
world an image of virtuous sobriety, discipline, and personal pro-
bity. Their architecture reflected this in its simple façades and large
windows (nothing to hide here), their clothes in severe colors, and
an absence of frivolous luxury. Even their graphic culture demonstrated their pride and self-image; witness Rembrandt's portraits of
his merchant contemporaries. The paintings and prints decorating
their houses bore hidden moral messages, but they also revealed a
world that married symbolism with a solid Dutch realism. Their
landscapes depicted local scenes without mythological nymphs,
fauns, gods, saints, or heroes; their seascapes dramatized the dangers of the element that had made them rich, preferably with scenes
of storms and shipwrecks; their portraits concentrated on faces and
hands, not on accessories or allegories.

Like truth and knowledge in the Republic of Letters, wealth and
poverty formed a moral universe that had still not quite escaped the
grip of theology, though the arguments and justifications were also
weapons in a conflict about social interests. Ideas of God's will and
His grace were still angrily and bloodily fought over by theologians
and indeed entire armies, and warfare was still the most common
recourse for the interpretation of social questions, or of the facts
of nature.

Here the Reformation had played a crucial role in opening up
new avenues of thought. God's will was no longer evident and
uncontested. Whatever one theologian said might be disputed
by another one, and this dissipation of rhetorical power had also
resulted in the fact that voices of a quite different kind were now
making themselves heard, voices speaking of nature in and of itself,
existing without necessarily being the manifestation of some higher
will. These voices belonged to scientists and philosophers, the
authors of anonymous pamphlets, and increasingly, the new urban

classes of educated people. As literacy rates continued to rise in the Netherlands, these voices were able to make themselves heard more and more clearly.

LEVIATHAN

For religious observers, declaring nature out of kilter provided an opportunity for interpreting natural events as divine portents. This intersection of physical events and frameworks of thinking made for some particularly interesting contrasts. The beaching of more than forty sperm whales on the Dutch coast—most likely because cooling seas had lured them into shallow waters in search of migrating prey—is a case in point.

Contemporary illustrations and pamphlets suggest how sensational and significant the deaths of these marine giants were judged to be. On February 3, 1598, the painter Hendrick Goltzius was inspired

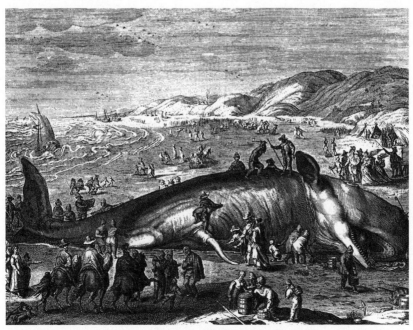

Divine sign or natural occurrence? A vigorous debate arose around the massed beaching of whales on the Dutch coast.

to draw one of them from nature, on the beach where it had been washed up, a sixty-five-foot (twenty-meter) bull whale. In an etching created by Goltzius's pupil Jacob Matham, small figures crowd around the creature, climbing it like mountaineers, or like the victors in a battle against nature itself. One of them is engaged in measuring the gigantic penis of the animal with a yardstick; others are already hacking into the body and carrying away the blubber in buckets.

The philosopher Hugo Grotius, who was a student in Leiden when the whale was found, recorded how clearly the reactions of his contemporaries were split, according to their degree of education. The event was "rare, but not miraculous, for real miracles do not exist," he wrote, adding:

> A fish from the species of whale . . . seventy foot long covered with its large body the entire space between the sea and the dyke. . . . And a large crowd had assembled in order to see this novelty, without fear of the bad smell emanating from the ruptured entrails of the fish so that the air was infected, causing sickness and even the death of several persons. It is still certain, though, that even those who only had mediocre judgment did not regard it as a miracle that a fish, which is in no way different from thousands of others hiding under the enormous extent of the oceans, had been hurled onto the sand by the waves. . . . But among the little people [*parmy le vulgaire*] . . . there were some who said that the death of the animal presaged a great victory with rich spoils, while others argued on the contrary that this adventure meant all kinds of ill fortune for those visited by the monster. And they were so convinced of their opinions that they even published them in print. But the vanity of these arguments was discovered during the remainder of the year, which was uneventful in comparison to the previous ones.[40]

The dry reaction of the famous legal thinker testifies to the growth of cultural differences: While simple people continued to believe in portents, even those with "mediocre judgment," who had only a little education and superficial knowledge, understood the beaching of the whale as part of the course of nature.

There were good reasons for the fascination sparked by these beached giants—not only for the crowds of everyday onlookers but also for the painters and scholars of the day. In the Netherlands, the Leviathan was a powerful symbol, almost the embodiment of the ongoing toils and triumphs of a nation living from the sea. The story of the prophet Jonah, who was swallowed by a whale for refusing to preach the Lord's truth, had powerful resonances in a country riven by religious wars, especially since parts of it had been prised from the sea itself by human ingenuity, causing some to fear that God might punish the arrogance of men daring to tamper with the order of Creation. The relationship between the Dutch and the North Sea was immensely strong—and ambivalent. The waves transported immense wealth aboard merchant ships and trawlers, but the power of the waves could also break the dams, inundate entire landscapes, and cast huge animals out of the depths. At the same time, the very position of the coast not only enabled the Netherlands to have access to the open sea but also made it vulnerable to attack from enemy fleets and devastating storms.

Wealth and poverty, mastery over nature and the danger of devastation, pride in civic achievements, and fear of hubris, fate, and divine retribution always coexisted in the Dutch mind. Now it was joined by the cognitive dissonance between divine portents and the course of nature. While the arguments continued on one side and the other, the nation's artists were attempting a reconciliation.

AN INVENTORY OF MORALITY

Music and the theater, in their high forms, were still largely courtly arts, celebrated in aristocratic settings. But in bourgeois Amsterdam, the premier form of artistic expression was painting, and the prevalent taste was not for the princely pomp of the Catholic Counter-Reformation. Atmospheric landscapes with dramatic clouds, seascapes with ships tossed by huge waves, dark and sober portraits, and winter landscapes dominated drawing rooms, but the most intimate and most intensely symbolic genre was that of smaller scenes: still lifes, in which an entire metaphorical universe was assembled in the guise of a few seemingly random objects. If the common people thought that a beached whale could be a divine portent, educated people also knew how to discover metaphysical significance in the seemingly commonplace.

A typical still life of the time, painted in 1668, shows a gray-marble table with a slightly chipped edge. Assembled on it are household items that appear to have been picked up carelessly and left there: a bunch of flowers, a carafe of aqua vita (brandy), letters, an accounts ledger and some books, a celestial globe, an hourglass, a ginger jar being used as an inkwell, a purse—and a skull. This glinting hint from a dead man's grin is almost unnecessary—for this is a painting about transience.

Maria van Oosterwijck: *Vanitas Still Life*.

The finely executed flowers, among them a gorgeous red-and-white tulip, will soon wilt, and even now the composition gives the impression that the flowers are sprouting from the bony hollow that once contained a brain. In the left corner, a head of wheat, symbol of the trading center's wealth, is being eyed by a mouse that will soon devour it, while the rattle, used by plague victims, is a reminder of the fragility of happiness. The recorder receives its voice only from living, human breath, which will soon expire. The cob of corn (a plant imported from South America) is already half eaten, and the ginger jar made from Chinese stoneware, which once contained a rare delicacy offered to select guests, has come down in the world to less exalted service.

The trading theme continues in the ledger, even if the astute observer already suspects that it contains not the bookkeeping of a wholesaler but rather a more existential tally, a register of sins. The book next to it, identified by its bookmark as *Self-stryt*, was a popular Protestant book of moralizing verse, first published in 1620. The celestial globe dominating the scene from the shadows represents not only the Creation but also the newfound human confidence in cartography and observation, in making the world, as it were, readable.

The carafe on the left is filled with a red liquid not at all like clear, distilled alcohol. What then is this aqua vita, this elixir of life, that is kept inside? Is it a liqueur or does the glass, fragile as the human body, contain a more precious liquid? The entire room is mirrored in the shiny, arched surface of the bulbous vessel, painted with discreet virtuosity into a space hardly larger than a fingernail: the large slatted windows and the trees outside, a sun-drenched wall, and, in the middle, almost invisible to the casual observer, a palette with colors arranged in a neat row, and above it the face of a woman, Maria van Oosterwijck.

Maria van Oosterwijck (1630–1693) was one of the few successful female painters of her time. Her still lifes were prized across the Continent. She had paid a price for her success, though, and for her independence: She never married, never had children. A deeply religious artist, she enlisted the entire visual repertoire of Protes-

Detail, carafe.

tant theology in her work, placing heavy emphasis on the theme of earthly transience. But while her paintings are meditations on mortality and contemporary religious mores, they also have a more immediate dimension. They show a society in which exotic objects such as Chinese stoneware, South American plants, and celestial globes had become a part of everyday bourgeois life—a world in which people read and kept books about gains and losses, in which engineers reclaimed land from God's sea (the author of *Self-stryt*, Jacob Cats, had become famous in the Netherlands for supervising the reconstruction of dams), but also a world in which people had become uncertain about how to understand their own era.

The artist's emblem is also part of the scene: a Red Admiral butterfly, which has found life after death in the chrysalis, like Christ himself. Its quivering, powdery wings show their beautiful pattern but will soon be dust again. For a moment, the butterfly is sitting on the book that contains the accounts of people's deeds and misdeeds. One moment later, it will be gone.

ON COMETS AND OTHER
CELESTIAL LIGHTS

Brandish your crystal tresses in the sky,
And with them scourge the bad revolting stars
That they have consented unto Henry's death!
King Henry the Fifth, too famous to live long!
England ne'er lost a king of so much worth.

SHAKESPEARE, *King Henry VI, Part I*

THE MADNESS OF CROWDS

On November 14, 1680, and again on December 24, a comet blazed a lonely, majestic trail across the European skies. From time immemorial, comets had been perceived as celestial signs, and many people looked up into the night full of fear. One of them wrote:

I tremble when I recall the terrible appearance it had on Saturday evening in the sky, when it was observed by everybody with inexpressible astonishment. It seemed as though the heavens were burning or as if the very air were on fire.[1]

The anonymous eyewitness had not exaggerated. The winter of 1680 was exceptionally cold, with clear, frosty nights—it was the apex of the Little Ice Age.

The comet that appeared during that winter was visible even during the day. In mid-December, at the height of its orbit around Earth, its tail extended across the horizon. The German theologian Johannes Bödiker was one of many observing the phenomenon. One year later, he published his *Christlicher Bericht von Cometen (Christian Report about Comets)*, in which he warned his readers:

> Because the great, wondrous god has once again moved the skies and has lit another comet, so the earth was moved in manifold ways, and humankind does not respond in the correct way: yea, while such miracles were seen by all eyes, they do not produce the same effect in all. For some go so far as to investigate out of mere curiosity, or they are so arrogant to prophesy what the comet will bring; many show too little response and react with mockery to the new miracle, as if it did not matter at all. Many are without all fear, not only of God's signs, but of the Lord himself.[2]

The Presbyterian Robert Law watched the celestial spectacle from his vantage point in Scotland:

> This comet, seen in December 1680, was seen on Friday, before the change of the moon. . . . [It] had a great blazing fra the root of it, was pointed as it came from the star, and then spread itself; was of a broad and large ascent up to the heavens . . . the stream of it amounted to our zenith, and beyond it, very terribly and wonderfully. No history ever made mention of the lyke comet, and it is doubted if ever the lyke was seen since the creation, and is certainly prodigious of great alterations, and of great judgements on these lands and nations for our sins; for never was the Lord more provoked by a people than by us in these lands, and that by persons of all ranks.[3]

The appearance of the comet terrified observers from Europe to the North American settlements, the Philippines, and China. In

A fiery messenger? The 1680 comet caused panic throughout Europe.

European cities, it caused a sensation fueled, like every good sensation, in equal parts by naked fear and by irresistible curiosity about everything new and strange. In his Paris observatory, Dutch astronomer Christiaan Huygens was practically besieged by questions from the anxious public. "It is beautiful today," he wrote to his father Constantijn, "and this brings countless people to the observatory, for people think that astronomers can explain and even interpret this phenomenon. Even I have been consulted two or three times."[4]

Huygens was a scientist, and he had no intention of being turned by the public into a necromancer. Like his colleagues Newton and Halley in Britain, he was interested in the comet as a physical phenomenon. Even as old wives' tales and prophesies were hawked on every street corner, often embellished with lurid details, he remained committed to plain observation.

Another scholar noted with surprise that the comet seemed to have the power to turn normally sensible people into uncontrollable hysterics looking to him for help. When the comet appeared in December, his quiet existence suddenly seemed to have come to an end:

As I was a professor of philosophy at Sedan [northwestern France] when there appeared a comet in the month of December, in the year sixteen hundred and eighty, I found myself incessantly exposed to the questions of several curious, or alarmed, persons. Insofar as I could, I reassured those who were bothered by this supposed bad presage; yet I gained but little by philosophical reasonings. The response was always made to me that God shows forth these great phenomena in order to give sinners time to ward off, by their penitence, the evils that hang over their heads. I therefore believed that it would be very pointless to reason further, unless I were to employ an argument making it manifest that the attributes of God do not permit him to intend comets to have such an effect.[5]

The philosopher who penned these lines had just turned thirty-three when the comet turned night into day, churning the fears and feelings of his contemporaries. His name was Pierre Bayle (1647–1706).

Bayle was the son of a Huguenot preacher in the Pyrenees. Born poor, he could only briefly attend school and was then taught at home until his older brother finished school—the family had not enough money to pay the fees for two boys. Only afterward could the nineteen-year-old Pierre continue his education—at one point, in a class with twelve-year-olds. Having grown up speaking Occitan, he learned French during his formal schooling.

Perhaps it was the lack of money and of a professional perspective that convinced him, two years later, to convert to Catholicism; he was then awarded a scholarship to study at a Jesuit college in Toulouse. But after two more years, he returned to his Protestant roots and had to flee Toulouse, labeled a heretic and traitor and threatened with severe punishments. He moved to Protestant Geneva, where he earned his keep as a private tutor.

When the philosophy professorship in Sedan became vacant, Bayle dared to return to his native France. The brilliant young scholar gained the post at the Protestant academy, but his quiet university life was soon upended once again. The town's Catholic

Tel fut l'illustre Bayle, Donneur des beaux esprits,
Dont l'élégante plume, en recherches fertile,
Fait douter qui des deux l'emporte en ses écrits,
De l'agréable ou de l'utile.

D. L. M.

L'Estampe se trouve chez J. Rollin fils quay des augustins à St Athanase.

A caprice of nature: Pierre Bayle argued against the idea of
divine intervention.

authorities forced him to flee once more, this time to the Netherlands, where he found employment as a teacher for the children of Huguenot refugees in Rotterdam. Finally he could devote his free time to his intellectual work without being threatened with persecution. His travel bag contained a manuscript with the mysterious title: *Lettre à M. L. A. D. C., Docteur de Sorbonne, où il est prouvé que les comètes ne sont point le presage d'aucun Malheur.*

Bayle settled in Rotterdam and would never again leave the city. Here his manuscript was published in 1682. The second edition came out with an altered title, under which the work is still known: *Pensées diverses, écrites à un docteur de Sorbonne, à l'occasion de la comète qui parut au mois de décembre 1680.*

As its educated readers would have expected, Bayle's *Various Thoughts* is full of rhetorical ruses and philosophical feints, which the author freely acknowledged in his preface to the second edition. Originally he had intended to publish the work in France. In order to evade both the strict censors and the threats to his person, he had been forced to pretend he was a Catholic and support Catholic positions. Now, in the Netherlands, this caution was no longer necessary, so instead of rewriting the entire book, he preferred to explain to his readers why the first edition had been written in such a coded manner—an indication that the Dutch reading public was used to more frank exchanges of ideas.

The text consists of a series of fictitious letters to a doctor at the Sorbonne, a description that would have conveyed to contemporary readers that the doctor was deeply dogmatic and staunchly conservative. In his own impeccably polite way, Bayle destroys his interlocutor with a few well-aimed stabs and slashes of his quill:

> I hear a number of persons reasoning every day on the nature of comets, and although I am not an astronomer in fact or by profession, I do not fail to study carefully all that the most clever have published on this matter; but I must admit to you, Monsieur, that none of it appears convincing to me, except what they say against the error of the people, who want comets to threaten the world with an infinite number of afflictions.
>
> It is this that makes me unable to understand how so great a doctor as you—who should be convinced, simply by having predicted correctly the return of our comet, that these are bodies subject to the ordinary laws of nature and not prodigies that follow no rule—has nevertheless allowed himself to be drawn along by the torrent and imagines, with the rest of the world,

despite the reasons of the select few, that comets are like the heralds of arms who come to declare war against mankind on behalf of God. If you were a preacher, I would pardon you for it because these kinds of thoughts, naturally very suited to be dressed in the most pompous and most pathetic ornaments of eloquence, do much more for the honor of him who pronounces them, and make much more of an impression on the consciousness of the listeners, than do a hundred other propositions demonstratively proved. But I cannot appreciate that a doctor, who has nothing to persuade the people of and who should nourish his mind only with an altogether pure reason, has such poorly supported sentiments in this matter and contents himself with tradition and passages from poets and historians.[6]

Comets as divine messengers? Bayle was amused by the idea. Scholars and poets who see portents in everything are simply hopelessly overoptimistic about their insights, "For in the end, one has to imagine that a man who has put himself into the spirit of making a poem has at that moment grasped the whole of nature. Heaven and Earth no longer act except by his order; eclipses and shipwrecks happen if it seems good to him; all the elements are moved as he finds appropriate. . . ."[7]

While poets can hardly be regarded as credible witnesses, historians tend to be worse. Most of them, Bayle is convinced, invent things in order to make their stories more attractive to their readers. After this opening salvo, Bayle begins to take apart the beliefs about comets piece by piece. There are many reasons to look at celestial bodies as purely physical phenomena, he writes: They are too small, too far away, their light is harmless, they do not produce a trail of poisonous vapors. If they were to have an effect on earthlings, it might just as well be a positive one: "there are misfortunes without comets and comets without misfortunes."[8]

The only thing believers are left with is the worst argument of all, writes Bayle: "there has never been anything more impertinent, anything more chimerical, than astrology, anything more ignominious to human nature, to the shame of which it will eternally be true to say that there were men deceitful enough to fool others under the pretext of knowing the things of heaven, and men stupid enough to give credence to them."[9]

It is clearly below Bayle's dignity to answer such ridiculous theories. To counter them would be to "shoot little birds with the arrows of Hercules." And even if it were true that bad events frequently occur after comet sightings, who could prove that they are causally connected? Is it not just the selective perception of the superstitious crowds looking for signs and oracles?

A true philosopher must not follow the principle of *vox populi, vox dei* (the voice of the people is the voice of god); he must seek to understand and interpret nature, which moves unerringly and without exception according to the laws decreed by God. It would make no sense to assume anything else. Why would the Lord in his goodness make the darkness of human behavior even more obscure by bad omens, violence, and catastrophes, instead of creating light? This superstition, argues Bayle, is un-Christian, even devilish, because it is the way of the eternal tempter to derail humanity from following God's truth.

The true domain of Satan is paganism and pagan rites, which always appear to insinuate themselves into the doctrines of the true faith. Immediately afterward, Bayle makes a characteristic volte-face: Atheists believe neither in God nor in the devil and therefore cannot be of help to Satan. Their erroneous belief system is therefore certainly not as dangerous as that of people who still worship the gods of antiquity, or those who see comets as heavenly signs, thus doubting God's omnipotence. Idolatry, not atheism, is the true enemy of faith.

From a reflection about natural phenomena, Bayle has suddenly moved into the heart of one of the mightiest controversies of his time: the battle between dogmatic religion and rational, scientific thought. Bayle pretends to take the side of the Church in judging this debate. Someone believing in an afterlife, he writes, is held to act morally.

On the other hand, someone who does not believe in predestination will laugh about these morals. Here Bayle knows that his imaginary correspondent and all his readers will be sympathizing with him, but this communion is short-lived, for Bayle next turns to hypocrisy and moral cowardice. If Christians act morally only in order to avoid punishment and to be rewarded—can this kind of action still be considered moral? Is this not more like the behavior of a gambler who pursues his own advantage?

Most Christians, Bayle writes, do not live by God's law. On the contrary, in their daily lives, people make case-by-case judgments, setting aside eternal principles. Christians do not live more morally than other people, he argues, quoting copious examples. Slowly, he moves in for the kill. A society consisting of atheists, he writes, would be in no way less virtuous than a Christian society:

> I will not hesitate to say that if one wants to know my conjecture concerning what a society of atheists would be like in regard to morals and civil actions, it would be very much like a society of pagans. It is true that very severe laws would be necessary there, ones very well executed with a view to the punishment of criminals. But are they not necessary everywhere? . . . One can say without being a ranter that human justice constitutes the virtue of the majority of the world, for when it relaxes the check on a given sin, few persons keep themselves from it.[10]

Human laws, not divine ones, cause us to act morally. For most people, living in accordance with God's will means nothing more than grandly forswearing sins they had no intention of committing, while at the same time continuing to live lives characterized by lechery, lies, stinginess, and greed. Someone who does not recognize the existence of God, on the other hand, can still live decently and charitably and control his cupidities, not because of the fear of eternal damnation but because he fears earthly punishments, and he also wants to gain the respect of others; this will guide his or her actions. Can a godless person be moral? Bayle parries this question with great ease. It is not more strange that an atheist leads a virtuous life than that a Christian commits all kinds of crimes, and, since his readers can see examples of the latter around them constantly— why should the former be so difficult to imagine?

After this dangerous excursion into religious questions, Bayle quickly returns to stargazing, natural laws, and the flawed logic of superstitions. He has made his point. Someone acting only from fear of punishment and hope of reward is a cool calculator, not a moral human being. An action can be called moral only if it contradicts a person's own immediate interests, as in the case of the heretic Vanini, in order to uphold a principle without gaining personal advantage. Atheists can do that just as well as saints.

Bayle's own principles were put to the test in 1685, a catastrophic year for the philosopher. In that year, Louis XIV of France revoked the Edict of Nantes, in which King Henri IV had granted at least limited rights to French Protestants. Fearful of a new wave of persecution, some 200,000 Huguenots fled—many to the Netherlands, others mainly to Geneva and Prussia, where the pragmatic elector welcomed the skilled and highly educated refugees with open arms. In his Rotterdam exile, Bayle realized that he would never be able to return home to France. Soon afterward, he received the message that his father and two brothers had died within weeks of each other—one of them, Jacob, in prison.

Pierre felt indirectly responsible for his brother's miserable end. He was the editor of a literary journal, *Nouvelles de la république des lettres*, which addressed itself to educated ladies and gentlemen all over Europe—in French. Bayle was careful not to print political or controversial material, and he limited himself to reviews and reports. At the same time, his journal was officially regarded as a Huguenot publication, and in spite of being banned in France, copies repeatedly found their way south, beyond the Dutch border. Pierre Bayle lived out of the reach of the French authorities, so they had simply arrested his brother Jacob, who was living in Bordeaux. Jacob could have regained his freedom by converting to Catholicism, but, like

Vanini and many other imprisoned Protestants, he had refused. He had survived only six months in prison.

Deeply saddened by this loss, Pierre Bayle also realized he would need to be more careful than before. He could not afford to be expelled from the Netherlands. He had wrapped and hidden his critique of religion inside a large amount of astronomical padding, but some Dutch guardians of the faith had become suspicious of the foreigner in their midst. He continued to publish, and he wrote tirelessly about religious tolerance and the futility of mutual hatred. In the context of the hate-filled debates of the day in which both sides claimed to possess God's truth, this was already an exposed position, and even if it did not lead to direct persecution, it made him the target of impassioned polemics in the Netherlands.

With great lucidity and conviction, Bayle demanded full freedom of conscience for all—not because it was the charitable thing to do, but because it was the only philosophically tenable position. His harshest critic was Pierre Jurieu, a former colleague from his days in Sedan, who had assisted Bayle in fleeing to the Netherlands and establishing himself in Rotterdam. Now he had turned into the philosopher's nemesis, and in 1693 he would even manage to have his former friend dismissed from his job as a schoolteacher, leaving Bayle penniless and without a regular income.

The precarious position in which Bayle now found himself was only to improve after he had completed a truly Herculean intellectual task. For years, he had invested almost all of his meager earnings in books. Now he managed to convince a bookseller (who also was a publisher) to pay him for writing a new and unique work. Bayle had set out to correct a Jesuit lexicon that to his mind was replete with inaccuracies, inconsistencies, and downright falsehoods. Soon, however, the project took on a dynamic of its own, eventually becoming a lexicon of historical, philosophical, and political personalities and ideas. After ten years of hard toil, the entire work—from "Aaron" to "Zuylichen"—finally could go to press. Toward the end of the writing process, Bayle was producing four reams of text, or thirty-two folio pages, every week; these were immediately sent to be typeset and printed.

At first glance, the result of Bayle's incredible effort, *Dictionnaire historique et critique*, in two (later three) volumes, hardly looks like a revolutionary document. Instead, it reads like a work of stupendous though somewhat pedantic scholarship. It was also full of curious editorial decisions, because at the speed at which Bayle forced himself to work, he had little time to be systematic. Still, few other seventeenth-century books have been so influential and so widely read and so frequently consulted as this *Dictionnaire*. Its enormous potential to change minds and shape debates, however, resided not so much in its content as in its method.

Bayle's *Dictionnaire* is a monumental encyclopedia of the philosophical, theological, and historical knowledge of his time: page after page of definitions, arguments, accounts, quotations. What makes it so crucially important for the debates of the subsequent decades and even centuries, however, was not his determination to correct the errors other authors had made, but rather the manner in which he set about it. All of the terms are first defined and then discussed at some length. Then these definitions are joined, undercut, contradicted, and subverted by whole choruses of footnotes and remarks, complete with full details of the printed sources—a practice that made the work a nightmare for the typesetters, who had to deal with pages of almost Talmudic graphic complexity.

Thanks to this method, Bayle did not have to include any controversial statements to ensure that attentive readers understood what he wanted to convey. Consider a random example, the article on the Greek philosopher Epicurus: The definition and the main text consist only of a few lines printed on the upper part of the page. Letters or numbers in brackets direct the reader's attention to literary sources or to footnotes.

Consulting these pages, the reader is inundated by heckling comments from doubters, explanatory notes, criticisms, and counterarguments by great authorities or obscure scholars. In addition to this, there are stories and anecdotes from the lives of great schol-

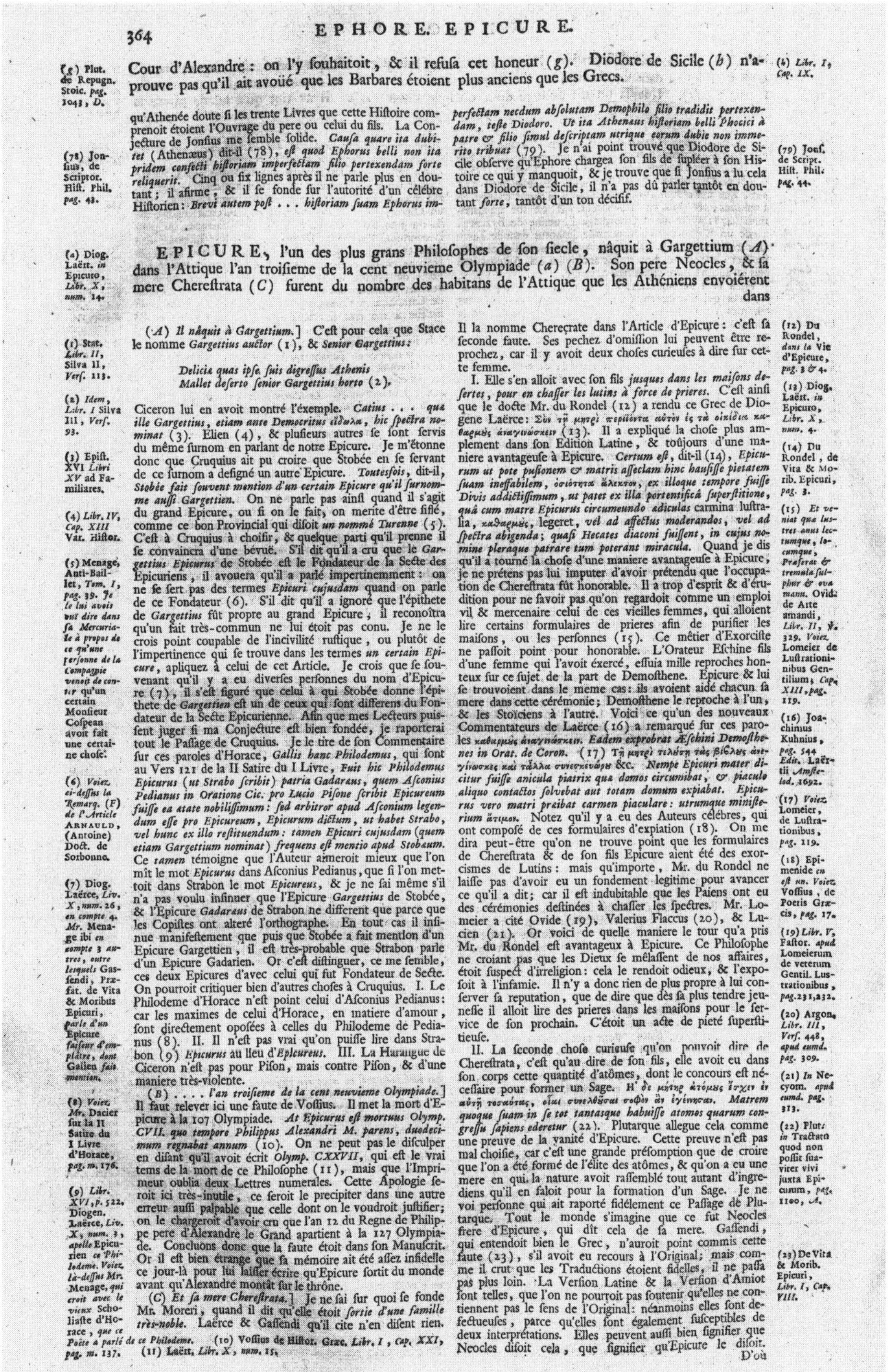
364 EPHORE. EPICURE.

Cour d'Alexandre : on l'y souhaitoit, & il refusa cet honeur (*g*). Diodore de Sicile (*b*) n'aprouve pas qu'il ait avoüé que les Barbares étoient plus anciens que les Grecs.

(*g*) Plut. de Repugn. Stoic. *pag.* 1043, *D.*

(*b*) *Libr. I, Cap. IX.*

qu'Athenée doute si les trente Livres que cette Histoire comprenoit étoient l'Ouvrage du pere ou celui du fils. La Conjecture de Jonsius me semble solide. *Causa quare ita dubitet* (Athenæus) dit-il (78), *est quod Ephorus belli non ita pridem confecti historiam imperfectam filio pertexendam forte reliquerit.* Cinq ou six lignes après il ne parle plus en doutant ; il afirme, & il se fonde sur l'autorité d'un célébre Historien : *Brevi autem post . . . historiam suam Ephorus imperfectam necdum absolutam Demophilo filio tradidit pertexendam, teste Diodoro. Ut ita Athenaus historiam belli Phocici à patre & filio simul descriptam utrique eorum dubie non immerito tribuat* (79). Je n'ai point trouvé que Diodore de Sicile observe qu'Ephore chargea son fils de supléer à son Histoire ce qui y manquoit, & je trouve que si Jonsius a lu cela dans Diodore de Sicile, il n'a pas dû parler tantôt en doutant *forte*, tantôt d'un ton décisif.

(78) Jonsius, de Scriptor. Hist. Phil. *pag.* 43.

(79) Jons. de Script. Hist. Phil. *pag.* 44.

EPICURE, l'un des plus grans Philosophes de son siecle, nâquit à Gargettium (*A*) dans l'Attique l'an troisieme de la cent neuvieme Olympiade (*a*) (*B*). Son pere Neocles, & sa mere Cherestrata (*C*) furent du nombre des habitans de l'Attique que les Athéniens envoiérent dans

(*a*) Diog. Laërt. *in* Epicuro, *Libr. X, num.* 14.

(*A*) *Il nâquit à Gargettium.*] C'est pour cela que Stace le nomme *Gargettius auctor* (1), & *Senior Gargettius :*

Delicia quas ipse suis digressus Athenis
Mallet deserto senior Gargettius horto (2).

Ciceron lui en avoit montré l'éxemple. *Catius . . . qua ille Gargettius, etiam ante Democritus εἴδωλα, hic spectra nominat* (3). Elien (4), & plusieurs autres se sont servis du même surnom en parlant de notre Epicure. Je m'étonne donc que Cruquius ait pu croire que Stobée en se servant de ce surnom a designé un autre Epicure. *Toutesfois*, dit-il, *Stobée fait souvent mention d'un certain Epicure qu'il surnomme aussi Gargettien.* On ne parle pas ainsi quand il s'agit du grand Epicure, ou si on le fait, on merite d'être siflé, comme ce bon Provincial qui disoit *un nommé Turenne* (5). C'est à Cruquius à choisir, & quelque parti qu'il prenne il se convaincra d'une bévuë. S'il dit qu'il a cru que le *Gargettius Epicurus* de Stobée est le Fondateur de la Secte des Epicuriens, il avouera qu'il a parlé impertinemment : on ne se sert pas des termes *Epicuri cujusdam* quand on parle de ce Fondateur (6). S'il dit qu'il a ignoré que l'épithete de *Gargettius* fût propre au grand Epicure, il reconoîtra qu'un fait très-commun ne lui étoit pas conu. Je ne le crois point coupable de l'incivilité rustique, ou plutôt de l'impertinence qui se trouve dans les termes *un certain Epicure*, apliquez à celui de cet Article. Je crois que se souvenant qu'il y a eu diverses personnes du nom d'Epicure (7), il s'est figuré que celui à qui Stobée donne l'épithete de *Gargettien* est un de ceux qui sont differens du Fondateur de la Secte Epicurienne. Afin que mes Lecteurs puissent juger si ma Conjecture est bien fondée, je raporterai tout le Passage de Cruquius. Je le tire de son Commentaire sur ces paroles d'Horace, *Gallis hanc Philodemus*, qui sont au Vers 121 de la II Satire du I Livre, *Fuit hic Philodemus Epicurus (ut Strabo scribit) patria Gadaraus, quem Asconius Pedianus in Oratione Cic. pro Lucio Pisone scribit Epicureum fuisse ea atate nobilissimum : sed arbitror apud Asconium legendum esse pro Epicureum, Epicurum dictum, ut habet Strabo, vel hunc ex illo restituendum : tamen Epicuri cujusdam (quem etiam Gargettium nominat) frequens est mentio apud Stobaum.* Ce *tamen* témoigne que l'Auteur aimeroit mieux que l'on mît le mot *Epicurus* dans Asconius Pedianus, que si l'on mettoit dans Strabon le mot *Epicureus*, & je ne sai même s'il n'a pas voulu insinuer que l'Epicure *Gargettius* de Stobée, & l'Epicure *Gadaraus* de Strabon ne different que parce que les Copistes ont alteré l'orthographe. En tout cas il insinue manifestement que puis que Stobée a fait mention d'un Epicure Gargettien, il est très-probable que Strabon parle d'un Epicure Gadarien. Or c'est distinguer, ce me semble, ces deux Epicures d'avec celui qui fut Fondateur de Secte. On pourroit critiquer bien d'autres choses à Cruquius. I. Le Philodeme d'Horace n'est point celui d'Asconius Pedianus : car les maximes de celui d'Horace, en matiere d'amour, sont directement oposées à celles du Philodeme de Pedianus (8). II. Il n'est pas vrai qu'on puisse lire dans Strabon (9) *Epicurus* au lieu d'*Epicureus*. III. La Harangue de Ciceron n'est pas pour Pison, mais contre Pison, & d'une maniere très-violente.

(*B*) . . . . *l'an troisieme de la cent neuvieme Olympiade.*] Il faut relever ici une faute de Vossius. Il met la mort d'Epicure à la 107 Olympiade. *At Epicurus est mortuus Olymp. CVII. quo tempore Philippus Alexandri M. parens, duodecimum regnabat annum* (10). On ne peut pas le disculper en disant qu'il avoit écrit *Olymp. CXXVII*, qui est le vrai tems de la mort de ce Philosophe (11), mais que l'Imprimeur oublia deux Lettres numerales. Cette Apologie seroit ici très-inutile, ce seroit le precipiter dans une autre erreur aussi palpable que celle dont on le voudroit justifier ; on le chargeroit d'avoir cru que l'an 12 du Regne de Philippe pere d'Alexandre le Grand apartient à la 127 Olympiade. Concluons donc que la faute étoit dans son Manuscrit. Or il est bien étrange que sa mémoire ait été assez infidelle ce jour-là pour lui laisser écrire qu'Epicure sortit du monde avant qu'Alexandre montât sur le thrône.

(*C*) *Et sa mere Cherestrata.*] Je ne sai sur quoi se fonde Mr. Moreri, quand il dit qu'elle étoit *sortie d'une famille très-noble.* Laërce & Gassendi qu'il cite n'en disent rien. Il la nomme Cherecrate dans l'Article d'Epicure : c'est sa seconde faute. Ses pechez d'omission lui peuvent être reprochez, car il y avoit deux choses curieuses à dire sur cette femme.

I. Elle s'en alloit avec son fils *jusques dans les maisons desertes, pour en chasser les lutins à force de prieres.* C'est ainsi que le docte Mr. du Rondel (12) a rendu ce Grec de Diogene Laërce : Σὺν τῇ μητρὶ περιϊόντα αὐτὸν ἐς τὰ οἰκίδια καθαρμοὺς ἀναγινώσκειν (13). Il a expliqué la chose plus amplement dans son Edition Latine, & toûjours d'une maniere avantageuse à Epicure. *Certum est*, dit-il (14), *Epicurum ut pote pusionem & matris asseclam hinc hausisse pietatem suam ineffabilem*, ὁσιότητα ἄλεκτον, *ex illoque tempore fuisse Divis addictissimum, ut patet ex illa portentificâ superstitione, quâ cum matre Epicurus circumeundo adiculas* carmina lustralia, καθαρμοὺς, legeret, *vel ad affectus moderandos, vel ad spectra abigenda ; quasi Hecates diaconi fuissent, in cujus nomine pleraque patrare tum poterant miracula.* Quand je dis qu'il a tourné la chose d'une maniere avantageuse à Epicure, je ne prétens pas lui imputer d'avoir prétendu que l'occupation de Cherestrata fût honorable. Il a trop d'esprit & d'érudition pour ne savoir pas qu'on regardoit comme un emploi vil & mercenaire celui de ces vieilles femmes, qui alloient lire certains formulaires de prieres afin de purifier les maisons, ou les personnes (15). Ce mêtier d'Exorciste ne passoit point pour honorable. L'Orateur Eschine fils d'une femme qui l'avoit éxercé, essuia mille reproches honteux sur ce sujet de la part de Demosthene. Epicure & lui se trouvoient dans le meme cas : ils avoient aidé chacun sa mere dans cette cérémonie ; Demosthene le reproche à l'un, & les Stoïciens à l'autre. Voici ce qu'un des nouveaux Commentateurs de Laërce (16) a remarqué sur ces paroles καθαρμοὺς ἀναγινώσκειν. *Eadem exprobrat Æschini Demosthenes in Orat. de Coron.* (17) Τῇ μητρὶ τελούσῃ τὰς βίβλους ἀνεγίνωσκες καὶ τἄλλα συνεσκευώρου &c. *Nempe Epicuri mater dicitur fuisse anicula piatrix qua domos circumibat, & piaculo aliquo contactos solvebat aut totam domum expiabat. Epicurus vero matri praibat carmen piaculare : utrumque ministerium* ἄτιμον. Notez qu'il y a eu des Auteurs célébres, qui ont composé de ces formulaires d'expiation (18). On me dira peut-être qu'on ne trouve point que les formulaires de Cherestrata & de son fils Epicure aient été des exorcismes de Lutins : mais qu'importe, Mr. du Rondel ne laisse pas d'avoir eu un fondement legitime pour avancer ce qu'il a dit ; car il est indubitable que les Paiens ont eu des cérémonies destinées à chasser les spectres. Mr. Lomeier a cité Ovide (19), Valerius Flaccus (20), & Lucien (21). Or voici de quelle maniere le tour qu'a pris Mr. du Rondel est avantageux à Epicure. Ce Philosophe ne croiant pas que les Dieux se mêlassent de nos affaires, étoit suspect d'irreligion : cela le rendoit odieux, & l'exposoit à l'infamie. Il n'y a donc rien de plus propre à lui conserver sa reputation, que de dire que dès sa plus tendre jeunesse il alloit lire des prieres dans les maisons pour le service de son prochain. C'étoit un acte de pieté superstitieuse.

II. La seconde chose curieuse qu'on pouvoit dire de Cherestrata, c'est qu'au dire de son fils, elle avoit eu dans son corps cette quantité d'atômes, dont le concours est nécessaire pour former un Sage. Ἡ δὲ μήτηρ ἀτόμους ἔσχεν ἐν αὑτῇ τοσαύτας, οἵας συνελθούσας σοφὸν ἂν ἐγέννησαν. *Matrem quoque suam in se tot tantasque habuisse atomos quarum congressu sapiens ederetur* (22). Plutarque allegue cela comme une preuve de la vanité d'Epicure. Cette preuve n'est pas mal choisie, car c'est une grande présomption que de croire que l'on a été formé de l'élite des atômes, & qu'on a eu une mere en qui la nature avoit rassemblé tout autant d'ingrediens qu'il en faloit pour la formation d'un Sage. Je ne voi personne qui ait raporté fidélement ce Passage de Plutarque. Tout le monde s'imagine que ce fut Neocles frere d'Epicure, qui dît cela de sa mere. Gassendi, qui entendoit bien le Grec, n'auroit point commis cette faute (23), s'il avoit eu recours à l'Original ; mais comme il crut que les Traductions étoient fidelles, il ne passa pas plus loin. La Version Latine & la Version d'Amiot sont telles, que l'on ne pourroit pas soutenir qu'elles ne contiennent pas le sens de l'Original : néanmoins elles sont defectueuses, parce qu'elles sont également susceptibles de deux interprétations. Elles peuvent aussi bien signifier que Neocles disoit cela, que signifier qu'Epicure le disoit.
D'où

(1) Stat. *Libr. II*, Silva II, *Vers.* 113.

(2) *Idem, Libr. I* Silva III, *Vers.* 93.

(3) Epist. XVI *Libri XV* ad Familiares.

(4) *Libr. IV, Cap. XIII* Var. Histor.

(5) Menage, Anti-Baillet, *Tom. I, pag.* 39. *Je le lui avois oui dire dans sa Mercuriale à propos de ce qu'une personne de la Compagnie venoit de conter* qu'un certain Monsieur Cospean avoit fait une certaine chose.

(6) *Voiez ci-dessus la Remarq.* (F) *de l'Article* ARNAULD, (Antoine) Doct. de Sorbonne.

(7) Diog. Laërce, *Liv. X, num.* 26, *en compte* 4. *Mr.* Menage ibi *en compte* 3 *autres, outre lesquels* Gassendi, Præfat. de Vita & Moribus Epicuri, *parle d'un* Epicure *faiseur d'emplâtre, dont* Galien *fait mention.*

(8) *Voiez Mr.* Dacier sur la II Satire du I Livre d'Horace, *pag. m.* 176.

(9) *Libr. XVI, p.* 522. Diogen. Laërce, *Liv. X, num.* 3, *apelle* Epicurien *ce Philodeme. Voiez là-dessus Mr.* Menage, *qui croit avec le vieux* Scholiaste d'Horace, *que ce Poëte a parlé de ce Philodeme. pag. m.* 137.

(10) Vossius de Histor. Græc. *Libr. I, Cap. XXI,*

(11) Laërt. *Libr. X, num.* 15.

(12) Du Rondel, *dans la* Vie d'Epicure, *pag.* 3 & 4.

(13) Diog. Laërt. *in* Epicuro, *Libr. X, num.* 4.

(14) Du Rondel, de Vita & Morib. Epicuri, *pag.* 3.

(15) *Et veniat qua lustret anus lectumque, locumque, Praferat & tremula sulphur & ova manu.* Ovid. de Arte amandi, *Libr. II,* ℣. 329. *Voiez* Lomeier de Lustrationibus Gentilium, *Cap. XIII, pag.* 119.

(16) Joachimus Kuhnius, *pag.* 544 *Edit.* Laërtii *Amstelod.* 1692.

(17) *Voiez* Lomeier, de Lustrationibus, *pag.* 119.

(18) Epimenide *en est un. Voiez* Vossius, de Poetis Græcis, *pag.* 17.

(19) *Libr. V,* Fastor. *apud* Lomeierum de veterum Gentil. Lustrationibus, *pag.* 231, 232.

(20) Argon. *Libr. III, Vers.* 448, *apud eumd. pag.* 309.

(21) *In* Necyom. *apud eumd. pag.* 313.

(22) Plut. *in* Tractatu quod non possit suaviter vivi juxta Epicurum, *pag.* 1100, *A.*

(23) De Vita & Morib. Epicuri, *Libr. I, Cap. VIII.*

The voice of reason: Bayle's *Dictionnaire* demonstrates on every page that the truth is always debatable.

ars, monarchs, and saints, often told from very personal perspectives that humanize these otherwise distant figures. Initially, the impression is one of confusion, but it soon becomes clearer. Bayle had collected (almost) everything that his era knew about these subjects—what had been known and discarded or refuted as well as what was currently being discussed. He was not interested in lexical definitions but rather in the living, ongoing debate.

Even the topic of Epicurus's place of birth turns out to be so complex that the page in question can fit only three lines of main text over a sea of footnotes. The first of these tackles opinions and quotations by Cicero, Horace, and lesser-known authors such as Asconius Pedianus, Strabo, and Pison, together with the author's own speculations on the question of whether sloppy or incompetent copyists were to blame for the fragmented and frequently contradictory information found in ancient manuscripts.

The following footnote deals with an error by a contemporary, Vossius, who recorded the wrong death date for Epicurus, possibly because the printer had left off two Roman numerals and the scholar had not noticed the omission. It goes on in this fashion, always with quotations, sources, and skeptical commentaries ("I don't know, on what Monsieur Moreri relies when he says that she [Epicurus's mother] had come from a very noble family"). Bayle builds a whole universe here, a Babylonian hubbub of scholarship, as he leads the way through the immense buzzing of voices ancient and modern.

As soon as we have moved past the biographical details, Bayle opens up the discussion of Epicurean ideas, at which point the footnote debate rises to another level. Ancient and modern authors are played off against one another, the opinions of critics and apologists are compared and quoted at length. A seemingly endless procession of names, quotations, and arguments moves past the reader's eye, seasoned by Bayle's own polite but persistent skepticism in a series of remarks of his own.

Like almost all philosophical texts of the period, the *Dictionnaire* can be read on several levels. While on the surface the definitions and discussions seemed authoritative, on another level it was almost immaterial what Bayle himself had to say about Epicurus, or indeed what others had thought about him. Since he brought so many voices into the general discussion, the ensuing dissonances created their own impressions and solidified into a crucial insight: Truth itself is not static, a thing one can own, but rather an open process. Those engaged in the search for it must be prepared to rethink their positions in the light of new evidence and stronger arguments. Knowledge is contested and contingent. It is not writ in stone, not on Mount Sinai and not elsewhere. It is no one's monopoly.

The *Dictionnaire*, all six million words on 3,300 folio pages, was published for the first time in 1697 and soon republished in an enlarged, three-volume edition. It became a bestseller of its era. Every great library, every Jesuit college, and every private collection of any standing had to have a copy of this monument of human erudition, methodical critique, and critical thinking. At the age of fifty, Bayle finally was sufficiently independent to quit his day job and devote himself entirely to scholarly research.

In spite of, or perhaps because of, its international success, the *Dictionnaire* caused further troubles for its intrepid author. The gentlemen of the Protestant Consistorium judged some of the passages scandalous, dirty, indecent, and obscene, and demanded omissions and alterations. Emboldened by his new position, Bayle reacted to this criticism not by making the changes but by adding a clarification that allowed him to make fun of the "dissatisfied" readers who found his work too true to life. It harks back to what he had written to his imaginary correspondent at the Sorbonne a decade earlier: "As a scholar of divinity. . . ."

Pierre Bayle died in Rotterdam in 1706, at the age of fifty-nine, after a long illness that would only have been exacerbated by his gargantuan workload, an illness that he appears to have treated with philosophy alone: "I have always refused all medical help. . . ."

The *Dictionnaire* was continued and updated after Bayle's death, and the last complete edition (except for modern reprints) appeared in 1821. Generations of pupils and students were taught by Bayle how to think, to use reason to get to the bottom of a question, to weigh arguments, and to live with contradictions and uncertainty, even in matters of religion. The philosopher himself did not describe himself as an atheist: "I am a good Protestant in two senses of the word," he once remarked, "because in my deepest soul I protest against everything people say, and everything that happens."

THE ANTICHRIST

The longest article in Bayle's *Dictionnaire* is also the most challenging. It deals with another philosopher who had lived in the Netherlands in precarious exile and who had died four years before Bayle had set foot in the country. As a thinker, this older philosopher had made enemies everywhere. His detractors simply called him the Antichrist, and his teachings were widely regarded as the most virulent intellectual poison ever produced by a human mind.

The source of this supposedly terrifying danger was an inconspicuous, polite, and soft-spoken man who earned his living by grinding lenses for scientific instruments: Baruch de Spinoza (1632–1677), son of a prominent merchant family in Amsterdam's tight-knit Portuguese Jewish community.

Bento, as the boy would have been known to his parents, grew up among people marked by the trauma of flight from Catholic persecution, forced conversions, and clandestine observances at the peril of their lives; the loss of their possessions, of family members who had been murdered or forced to flee elsewhere, of the country that had been home to them for centuries. In their new place of residence, the Jews not only were allowed to live and work, they also were free to live according to their faith. For those who had come from Portugal, Amsterdam provided the first breath of freedom for generations.

Bento was intellectually precocious and attended the yeshiva, or Talmudic school, at the majestic local synagogue close to the center of town, a monument to Amsterdam's tolerant treatment of its new Jewish citizens. On the streets of the city, the boy encountered

Thinking God out of the world: Spinoza's theological
meditations left no room for anything outside or beyond
nature and its immutable laws.

a world very different from his former home. Within less than a
century—since the beginning of the Little Ice Age and its rise as
a trading hub—the city had risen from the provincial backwater
where Wouter Jacobszoon had sought refuge, to a cosmopolitan
center, a gate to continents beyond the horizon, to vast fortunes and
opportunities, and to the most radical ideas and debates anywhere
on the Continent.

For some, the encounter between a traumatized community
seeking to recapture its traditional identity and the secular, foreign world around it led to tragedy, as the biographies of several
Amsterdam Jews demonstrate. One such case of conflicted iden-

tity, and exclusion from the community, had profoundly disturbed the Amsterdam Jewish community during Bento's childhood: Uriel da Costa (1585–1640) had made many epic journeys during his life. Born in Porto into a Jewish family that had been forced to convert, he had been raised as a pious Catholic, had studied theology at a Jesuit college, and had received lower clerical orders.

Da Costa seemed destined for a brilliant career in the Church, but his studies had set him on a different path. Secretly, he attempted to reconstruct from the Latin Bible how his own Jewish ancestors had lived and practiced their rituals during the era of the Temple in Jerusalem. Da Costa decided he wanted to live the life God had ordered his chosen people to live, and finally he renounced his Catholic vows, married a Jewish woman, and moved to the countryside. Three years later, in 1614, the couple decided to emigrate. They embarked first for Amsterdam and from there went on to Hamburg, where another Portuguese community had been established.

Finally free to practice his faith, the former cleric now faced, for the first time in his life, the myriad customs, laws, and bylaws that make up traditional Jewish life. But his theological training, with its (somewhat ironical) emphasis on rationality, had left its mark on his keen mind. He soon began to argue about laws he deemed redundant or even objectionable, especially if they were decreed not by the Bible itself but by the rabbinical Talmud. In order to make his case comprehensively, he penned a polemic with the inflammatory title *Propostas contra a tradição* (*Theses against Tradition*, 1616); this led to his being banned from the Hamburg Jewish community. Da Costa remained in the city with his wife, but when she died suddenly in 1622, he moved back to Amsterdam, where he hoped to find more sympathy for his critical positions, as well as a connection with the people he regarded as his own.

But da Costa's reputation as a religious agent provocateur had preceded him, and he was soon made to realize that he was not welcome in Amsterdam. The following year, he was formally excluded from the city's Jewish community. Determined to justify himself, he published a second book, in which he not only rejected the authority of the Talmud but also argued that the Bible does not contain the idea of the immortality of the soul, and that human souls are therefore mortal. That was too much even for the Calvinist city fathers, who ordered the book publicly burned and exiled its author to Utrecht.

Although Utrecht was a prosperous and pretty town, it could not answer da Costa's restless search for belonging and identity. Little is known about his years there, but it seems certain that he was desperately lonely. Time and again, he attempted to make contact with the Jewish community in Amsterdam. In 1629, he was readmitted and moved back to the city, only to be excluded again four years later, after he had reportedly been caught several times breaking the dietary laws. Poor and without support, he had settled close to the Jewish neighborhood.

In 1639, da Costa's despair seems to have reached such a pitch that he was ready to renounce the error of his ways and return to the faith of his fathers. The rabbinical court condemned him to thirty-nine lashes, after which punishment he was forced to lie on the threshold of the synagogue and allow all members of the community to walk over his prostrate body. Although he subjected himself to this, it appears that the humiliation of it was ultimately too great for him to bear. In April of the following year, he shot himself to death with a pistol. He left behind an autobiography, *Exemplar humanae vitae* (*An Example of Human Life*, first published in 1687), in which he described his intellectual and religious path. It was his final justification for his life and thoughts.

When da Costa was whipped and humiliated in the synagogue, Bento Spinoza was only eight years old. Even if the boy was not present during the punishment itself, he certainly knew about it and understood the warning it sent to anyone considering exploring beyond the limits of what the community allowed. But while the elders could be harsh to individuals whom they regarded as apostates, they were not entirely closed to the world around them. Other members of the community, perhaps less headstrong and less openly challenging than da Costa, were able to move between inside and outside, between ancestral Jewish customs and new Dutch ideas.

One of the most important influences on Spinoza's intellectual formation was the Jewish scholar Menasse Ben Israel (1604–1657), leader of the Amsterdam yeshiva. He was a friend of intellectuals such as Hugo Grotius, the legal scholar who had written in amused tones about the beached whale on the Amsterdam shore. Ben Israel was also a friend of Rembrandt, who created an etching of him. He authored several mystical texts, and he published books in Hebrew, Yiddish, Latin, Spanish, Portuguese, Dutch, and English. His pupils were educated to be open toward the world. Bento, the most brilliant among them, absorbed all this, and soon he began a course of study that would bring him into conflict with the Jewish elders and eventually alienate him from his Jewish world.

This alienation from the life of the Jewish community, its traditions, and its ways of thinking developed over a number of years, and it was accelerated simply by bad luck. Bento's father's import business for wine, olive oil, dried fruit, and nuts had been suffering, due to the maritime blockades imposed during the first Anglo-Dutch War. After the deaths of his father and his brother Isaac, Bento, then twenty-two, was forced to abandon his books and take over the family firm, but he proved unable to rescue the now-struggling business—perhaps he was more interested in philosophy than in wine or dried fruit.

Baruch de Spinoza was in any case already pursuing interests beyond the physical and intellectual confines of the Jewish community. At almost twenty years of age, he had begun to study Latin, the language that would open up the entire canon of Western philosophy. He attended a Latin school on the Singel (the city's innermost canal). The school director, Franciscus van den Enden (1602–1674), was a typical product of the Dutch revolution in education. His own father, a weaver by trade, had made it possible for him to receive an education in his native Antwerp, and the young man had joined the Jesuits. After leaving the order, he had finally settled in Amsterdam, where he had opened an art dealership that also brought him in contact with the city's foremost artists. When his business went bankrupt, van den Enden had founded the Latin school; here the young Spinoza learned to read and write the lin-gua franca of scholarship. Here Baruch (or Bento) was known by the Latin name of Benedictus.

Van den Enden, who was immensely erudite, had left the Jesu-its because he had found that his own ideas about the world had begun to differ too much from theirs. But he had adopted many of the Jesuits' educational ideas, from which he himself had benefited. He performed Latin plays with his pupils, in which his own daugh-ters took the female leads, and he introduced his charges to secular authors such as Terence and Seneca. He also wrote two tracts about tolerance and about the possibility of an ideal republic, in which he argued for a secular state with democracy, free and equal education for males and females, a humane justice system, and the abolition of slavery.

While van den Enden was a good educator, it appears that his real passion lay elsewhere. His subsequent life illustrates his strength of character and perhaps, too, his naiveté amid the perils of seventeenth-century intellectual inquiry. Determined to see his political ideas implemented and to further the cause of justice in the world, he planned to establish an independent republic in Nor-mandy in northern France, which would be founded after an armed coup. It was to be the first revolution in Western history founded on secular principles. Van den Enden moved to Paris to pursue his revolutionary goals, and there he saw his plans gradually dissipate, eventually finding himself drawn into a conspiracy against Louis XIV. He was still hoping, however, to see the foundation of an ideal republic, for which he and his coconspirators had already drafted a constitution. But the plotters were betrayed, and, after the custom-ary interrogations and torture, the coconspirators were executed in front of the Bastille on November 27, 1674. As a commoner, van den Enden was hanged.

Attending van den Enden's school two decades before these events were to unfold, Baruch de Spinoza encountered a world of new ideas and perspectives among his fellow pupils, many of whom were the sons of French immigrants and were familiar with thinkers such as Descartes. Judging from his own later works, the young man was deeply impressed by the rigor and daring of the Cartesian method, and by its claim to scientific objectivity.

As these new intellectual horizons were opening up for him, Spinoza found it increasingly difficult to respect and observe the old religious laws that he had begun to regard as meaningless. But the members of Amsterdam's Jewish community were not only his trading partners—they were also his cousins, brothers, sisters-in-law, school friends, and neighbors. He knew that he would soon be expected to choose a girl from a respectable Jewish family and settle down to establish a family of his own. His new opinions and interests were met with deep suspicion in the community.

Friends and acquaintances unanimously described Spinoza's character as free of guile or aggression, and founded on strong personal principles; this apparently made it impossible for him to keep his opinions to himself, or to disguise his way of thinking once he had arrived at an idea through rigorous analysis. The young man who consorted with non-Jewish atheists and who devoured Latin books was constantly engaging others of his community in intellectual debates, embarrassing the rabbis with unusual, searching questions. Instructed to abandon his heretical ideas, Spinoza refused, and the community reacted harshly.

In 1656, the religious court of the Amsterdam community pronounced a Herem, a comprehensive ban that forbade all Jews, even members of his own family, to have any kind of contact or communication with the young man. Containing biblical curses and blood-curdling threats, this ban was the most virulent document of its kind that had ever been written within the community. Its wording indicates that Baruch's new way of thinking had been causing problems for a considerable time, and that he had attracted the ire and even the hatred of his elders:

The Lords of the *ma'amad*, having long known of the evil opinions and acts of Baruch de Espinoza, have endeavored by various means and promises, to turn him from his evil ways. But having failed to make him mend his wicked ways, and, on the contrary, daily receiving more and more serious information about the abominable heresies which he practiced and taught and about his monstrous deeds, and having for this numerous trustworthy witnesses who have deposed and borne witness to this effect in the presence of the said Espinoza, they became convinced of the truth of this matter; and after all of this has been investigated in the presence of the honorable *chachamim* [sages], they have decided, with their consent, that the said Espinoza should be excommunicated and expelled from the people of Israel. By decree of the angels and by the command of the holy men, we excommunicate, expel, curse and damn Baruch de Espinoza, with the consent of God, Blessed be He, and with the consent of all the holy congregation, and in front of these holy scrolls with the 613 precepts which are written therein; cursing him with the excommunication with which Joshua banned Jericho and with the curse which Elisha cursed the boys and with all the castigations that are written in the Book of the Law. Cursed be he by day and cursed be he by night; cursed be he when he lies down and cursed be he when he rises up. Cursed be he when he goes out and cursed be he when he comes in.[11]

As a merchant relying on a network of Jewish partners, Spinoza was finished, even before he had published a single line of his supposed heresies. He accepted the break with the religion of his birth and with the people among whom he had grown up. In contrast to the unhappy Uriel da Costa, he never made any attempt at reconciliation or return. Instead, he moved to the town of Rijnsburg, not far from Leiden, where he attended Leiden University for a while. Bereft of income and support, he learned the craft of lens grinding, a quiet and contemplative occupation that had become economically important after the invention of telescopes and microscopes half a century earlier.

Spinoza's chosen work of grinding lenses is almost irresistibly symbolic, but it also ruined his health. Perhaps he was already suffering from tuberculosis when he set out on his new life, but the glass particles that he regularly inhaled surely helped to intensify this and to accelerate his end. In spite of his illness, however, he made the most of his newfound freedom and began writing a philosophical work that became a topic of fervent discussion long before its publication.

The ideas that so excited the freethinkers and philosophers in Spinoza's orbit were controversial and dangerous. Pierre Bayle was to write, in his *Pensées diverses,* that they were based on "the most monstrous hypothesis imaginable, the most absurd, which is most opposed to all concepts evident to our mind."[12] Always a master of disguises, Bayle tucked away in a footnote his elucidation of this monstrous idea.

Spinoza's great work, *Ethica (Ethics)*, was published in 1677, the year of his death at the age of forty-four. The first part is devoted not to human beings and their actions but to the exact nature of God, from which everything comes and in whom everything is contained. These thoughts are developed in numbered theses and proofs, conclusions and examples, after the geometrical method; Spinoza was reasoning like a mathematician or a scientist.

God is everything: infinite, indivisible, and omnipresent, the uncaused cause of all being. He is the substance of the world. Since there can be no two substances (in which case, God would be limited), it follows that everything existing in the world is part of God's substance, that God is everything we can perceive, think, or feel. Our experience consists of the experience of God's attributes in time and space, and our reason, the faculty closest to God's own being, allows us to understand his Creation.

Thoughts such as these could have been formulated by any number of theologians of Spinoza's day, but he took them to their logical conclusion, pushing them farther than anyone had done previously. If nature is nothing else than the perceptible expression of God's attributes, and if God is perfect, then it must follow that nature itself was perfect and that nothing in its eternal course could be altered or changed, because it would alter the nature of the Creator himself—and to alter something perfect would mean to make it imperfect. Indeed, not even God himself can alter the course of nature, because he cannot alter his own divine nature.

Deus sive natura (God or nature) was the shorthand for this idea, as Spinoza himself remarked in the Latin introduction: God and nature became synonyms, two sides of the same coin, two words for the same all-encompassing reality.

Having started from a classic, almost orthodox theological standpoint—God is omnipotent, omnipresent, perfect, the substance of the world—Spinoza suddenly topples his reader into the abyss. If God is synonymous with nature and her immutable laws, he cannot also be the God of the Revelation, of the covenant and of redemption, and his name is nothing more than a mythologically charged synonym for the world around us. That means that praying is meaningless, because even if he wanted to, God could not change the course of nature, and anyway, if everything is part of God and therefore perfect and necessary, so is suffering.

Miracles, too, are impossible and have never occurred. The stories about them are either exaggerated for poetic effect or simply based on faulty human memories, all too human—just like the prophets, who had no special line of communication to the Creator but were simply speaking from their own understanding. Rituals, then, are useful only as methods of social control, heaven and hell are old wives' tales, fanaticism and the Inquisition are barbaric, and power by the grace of God is nothing but the convenient invention of those wielding power on Earth. It is not difficult to understand why Spinoza chose to publish his work anonymously.

Spinoza's arguments derive their extraordinary power not only from their logical force but also from the almost Olympian calm with which they are developed. Where others polemicized and agitated, he simply unfolded his argument step by compelling step. This, however, was only the argumentative groundwork enabling the philosopher to shift his attention to human beings and moral actions. Having identified God with nature, Spinoza elaborated an ethics of living morally within and through nature. Truly moral actions cannot be found where people simply follow laws to avoid punishment or seek reward, especially if these laws are of human origin. Whoever is led by passion and desire cannot be free. Freedom can be gained only through insight and through rational control of the passions—not because passions are sinful in themselves, as the Christians and others believed, but because they entail suffering and shackle the mind. This insight revealed what were for Spinoza the true and eternal principles of ethics: Morality lies in doing good simply because it is good.

Everything that is in accordance with nature is good. We tend to act out of self-interest, from egotistical motives, driven by our will to survive. However, if we analyze our situation, it becomes clear that our best chance to survive well, and with the least degree of restraint, lies in acting in solidarity with others in order to create a world in which people can live with dignity.

It will be obvious why Spinoza's ideas provoked such unbridled hatred. He never argued against the existence, omnipotence, or benevolence of God; on the contrary, he made God's perfection the cornerstone of his entire ethical edifice. But starting from there, he shattered religious beliefs and rituals, calling them man-made superstitions. If God and nature were the same, the only rational response was to live in accordance with the laws of nature and to cultivate empathy and solidarity among all humans, all living beings. The driving force of this ethical behavior was life itself, was desire, self-interest. Aided and fully realized by a rational mind, this in itself was sufficient grounds for morality, and for the emergence of a just and tolerant Enlightened society.

Spinoza had done nothing less than find an exit from the intellectual cul-de-sac in which the Western intellectual tradition had been trapped for centuries. Before Spinoza, methodical thinking about ethics in the Christian tradition had taken place in a strictly confessional framework, constantly seeking affirmation from the Bible or from other theological authorities. There had been a tradition of secular thought (we have encountered several of its exponents), but this had been generally directed against religion, and it therefore

had to play by the rules of the theological discussion. Secular thinkers had also been persecuted, forced into exile, or even killed, thus
preventing their ideas from entering the main debate. Now a humble lens grinder living in the Dutch provinces upended this entire
debate by using the language of theology to build a secular ethics.

It is tempting to reflect on whether Spinoza would have cast his
universalist ethics in such a theological mold if he had been writing
two or three centuries later. In his own time, however, it was exactly
this language that gave his ideas such immense power. "Benedictus" anchors his argument in the safe harbor of theology, but as
soon as his passengers are on board, he lifts anchor and sails off
toward new horizons.

Drawing from scholastic theology and its offspring, rationalist
philosophy, as well as from Jewish thought, Spinoza was able to
construct this argument precisely because he had encountered both
traditions and was living a life without religious or institutional
alignment, without patron or stipend, thus allowing him to maintain a detached perspective on the great debates of his time. Ideas,
after all, are not born on the pages of books but in the minds of
people living in specific biographical and historical situations, with
specific mental horizons, speaking a language that is particular to
their time and place.

The two traditions in which Spinoza had been educated joined
forces in his argument. Initially, he appears to use Descartes only as
a formal inspiration, and to use medieval scholastic theology as the
bedrock on which the entire edifice is built. God's indivisibility and
oneness, his necessary perfection, his omnipresence and omnipotence and benevolence, man (and woman) as his creatures partak-
ing in the divine light through reason—all this was the teaching of
the scholastic master thinker Thomas Aquinas, while substance and
attributes are concepts coming directly from Aristotle. This was the
tradition the young man had encountered in the Latin school and in
the lessons of his ex-Jesuit teacher Franciscus van den Enden—the
world in which he was known as Benedictus.

The greatness of his philosophical enterprise consisted in fusing this scholastic perspective with that of Baruch, the former Talmud scholar. Each tradition offered questions, techniques, and perspectives that the other lacked. As a thinker living with and between different traditions, languages, and attachments, Spinoza was in an ideal position to enlist the fruitful tensions arising between them.

The Jewish tradition differs from its Christian counterpart in that it is less theological—i.e., less focused on God and his attributes— than legal, occupying itself with specific cases, rituals, and practices. The rabbis of the Talmud had never been interested in such questions as, How many angels can dance on the head of a pin? They asked questions of a more practical nature: What constitutes work on the Sabbath? When is meat kosher? If a stone wall in an interior courtyard collapses, which of the neighbors is required to pay for the repairs, and to whom do the stones belong?

For centuries, exiled Jewish communities had been able to maintain their culture and identity, because their holy of holies was not the destroyed temple in Jerusalem but a book that they could carry with them, whose wisdom could be unlocked only by painstaking interpretation and vigorous debate. These debates were printed on the pages of the Talmud in layer upon layer of discussion, commentary, dissenting opinions, and explications, all accrued over centuries like layers of soil on holy ground. Those studying the Talmud by reading it aloud and talking to one another about arguments, hidden meanings, and logical proof became part of this intergenerational conversation. Studying the Talmud and the resulting complexities of ritual observance became the central pillars of Jewish identity.

The central axiom of this chain of common discussions was the literal truth of the Bible. Even apparent spelling mistakes in the Torah scrolls were handed down from scribe to scribe, on the assumption that the smallest dot must have been divinely inspired, and that resolving apparent problems might be the first step to more profound revelations. This had stabilized the Jewish tradition over the centuries, but it also confronted communities with a dilemma. The Torah described a Bronze Age society, its laws and morals. Much of what was described had been otherwise forgotten, and even the meanings of words were the subject of intense debate among Talmudic scholars. The attitudes and social practices of the Israelites of biblical times were at odds with the social realities and intellectual horizons of later Jewish communities living in Babylonian exile, in Moorish Spain, in medieval France, or in the shtetls of Galicia. Their moral ideas had changed from divinely sanctioned genocidal Bronze Age invasions, public executions, and violence in the name of the Lord, to more abstract ideas of justice and truth.

Jewish diaspora communities faced a common problem: How could they reconcile the centrality of the Torah and its unchanging, literal truth with lives led according to palpably different moral standards? One strategy was simply to argue that laws such as capital punishment could only be applied in Israel, and in the presence of the Temple in Jerusalem, long since destroyed. Another approach, however, proved infinitely more subtle and effective. The secret was to take the divinely inspired letters of the holy scriptures literally—as literally as only divinely inspired documents should be taken. In them, there is no coincidence and no mistake, but rather a vast, mostly invisible net of hidden correspondences, mystical truths, undiscovered meanings, and layered insights.

Even if the rabbis of antiquity may not have intended it this way, later teachers had discovered a way of unlocking and deconstructing the holy texts that allowed them to be reassembled in any way they pleased. By reading the Bible not as a historical document but as written revelation, they could apply particular rules of interpretation that changed meanings altogether, thus bringing them more closely into line with behavior suited to their current lives.

These interpretative rules thus became keys to a certain intellectual freedom and moral self-determination. One verse could be explained by quoting another; hidden meanings could be discovered in errors of transcription; numerical values could be assigned to letters, so that words could be treated as numbers—added and subtracted, divided and multiplied, and replaced by words of equal numerical value, but with a totally different meaning once reconverted into letters. Once this kind of creative interpretation was unleashed, any text could be found to have a multiplicity of divinely sanctioned meanings.

To appreciate the perspective Spinoza would bring to the Christian scholastic tradition, it may be helpful to look at an exemplary Talmudic debate about a troublesome biblical verse, a famous passage from the tract Sanhedrin of the Babylonian Talmud, one of many taught to boys like Baruch Spinoza in his Amsterdam Jewish school—the law concerning the rebellious son.

Even for the scholars of antiquity, the fifth book of the Torah revealed a patriarchal Bronze Age in all its cruelty—so much so that subsequent generations found themselves hard-pressed to justify it. The twenty-first chapter deals with the question of when a man may take the beautiful wife of a vanquished enemy as a sex slave when he lusts after her; the following chapter decrees that a man who rapes a virgin must pay a fine to her father and marry her, while an unmarried virgin who chooses to sleep with any man out of wedlock is to be stoned to death. Flanked by these examples of divine wisdom is the law that proved such a challenge to later interpreters (Deuteronomy 21:18–21):

> If a man have a stubborn and rebellious son, which will not obey the voice of his father, or the voice of his mother, and that, when they have chastened him, will not hearken unto them: Then shall his father and his mother lay hold on him, and bring him out unto the elders of his city, and unto the gate of his place; And they shall say unto the elders of his city, This our son is stubborn and rebellious, he will not obey our voice; he is a glutton, and a drunkard. And all the men of his city shall stone him with stones, that he die: so shalt thou put evil away from among you; and all Israel shall hear, and fear.

Can it ever be justified to execute a child for disobedience? The Babylonian Talmud engages in a heated debate about these verses. Not once is the implicit truth and divine inspiration of the biblical text called into question by the discussants, who only appear to be seeking clarity as to the exact circumstances under which such a harsh sentence can be carried out.

In this spirit of textual clarification, the rabbis proceed to define what the text is actually intended to convey. A disobedient son obviously cannot yet be an adult according to religious law, because his behavior would no longer be his parents' responsibility, and he therefore must be younger than thirteen, the age at which his Bar Mitzvah ceremony marks his religious adulthood. But he must also no longer be a child who cannot be held responsible for his actions. How then is this difference to be determined? He must already be capable of fathering a child of his own. But how can this be known? Is it enough when two pubic hairs have appeared?

The next question relates to the definitions of the words *glutton* and *drunkard*. How much must one eat and drink to qualify? After intensive debate and consultation of other, unrelated Bible verses, the scholars come to the conclusion that the boy would have to down at least one barrel of Italian wine (which they regarded as the strongest of all) and must eat more than a kilo of raw meat a day. The meat also cannot be kosher, and even then it would only count if he wolfed down his giant helping in the company of "totally worthless men." As to the parents, the Bible clearly states that they must be in prime health and free from all infirmity (otherwise they could not "lay hold of him"), and when they say that their son does not "obey our voice," it follows that they must speak the exact same words at the same pitch and with the same intonation and melody.

The restrictions continue to rain into the arid Bronze Age scenario until the literal reading of the text has arrived at the obvious conclusion, as Rabbi Jehuda is quoted to have said at the end of the debate: "There never has been a rebellious son and there never will be one. Why was the law written? So that you can learn it and reap your reward." Close reading has transformed a barbaric law into an example of dedicated learning and principled morality. Close reading, as the rabbis had discovered centuries earlier, is a powerful tool.

It is this close reading of the Talmud that Spinoza applied to the thinking of the Christian tradition. He does not refute the arguments of scholastic theology, he simply takes them so seriously that they collapse under the weight of the expectations invested in them. Is God's substance infinite? Then God must be in everything, everything in God, then God is nature. Are his laws necessary and eternal? Then he himself must be bound by them and therefore unable to work miracles or to intervene in the course of history. God is nature, its laws are His laws, and nature is good. Human intellect is an expression of the divine mind, approaching God means understanding nature, and living according to His will means living according to the laws arising from the analysis of human nature.

It was Spinoza's position as an outsider that allowed him to use Christian theology against itself and to arrive in a purely material universe by reaffirming God's greatness. This made him dangerous. He had abolished the binary opposition between religion and freethinking and had shown a way out of the maze in which European philosophy appeared to be trapped. In his work, his readers saw a new pathway of thinking, leading to new, secular horizons.

Attentive readers will have noticed that Pierre Bayle, who wrote so harshly about Spinoza, defended philosophical positions bearing a remarkable similarity to those developed in the *Ethica*, which Bayle had very probably adopted from Spinoza. The necessity and rationality of tolerance, the rejection of natural phenomena as miraculous manifestations of God's will, the human origin of religious rituals—these themes are central to both philosophers. Perhaps Bayle's very orthodox ire against Spinoza is less surprising if one views it as a defense of himself: The members of the Calvinist congregation in Amsterdam were eyeing Bayle's *Dictionnaire* with great suspicion and were looking for an opportunity to silence him.

And Bayle was also making use of a tried-and-tested stratagem: In order to reject the philosophy of his rationalist, critical predecessor, he quoted his works frequently and at length, marshaled regiments of sources and battalions of refutations in the footnotes, and explained the heretical ideas he claimed to abominate and disprove. In doing so, he became one of the main sources of information about the thinker so many theologians regarded as the devil incarnate.

Even in the Netherlands, Spinoza's works were long banned and remained extremely difficult to find. Rare manuscript copies circulated from hand to hand, but these could not satisfy the curiosity his philosophy had kindled in countless minds. Bayle's loquacious and loving refutation of Spinoza's ideas not only satisfied all requirements of religious orthodoxy—it also became the single most important source of information about Spinoza's thought. Generations of readers turned to Bayle's withering criticisms to learn about the author whom so many authorities had described so damningly.

There was rare unanimity among the different churches when it came to banning and condemning the *Ethica* and the forbiddingly entitled, posthumous *Tractatus Logico-Philosophicus* and their author. The Jewish community had already expelled Spinoza and deemed his works heretical. The Protestant theologian Franz Buddeus called him "the prince of atheists of this time."[13] The Dutch professor Regnier Mansvelt declared the book to be damaging to all religions and worthy of being "buried by eternal forgetting," and the merchant Willem von Bijlenburgh opined that "this atheist book is full of abominations . . . which must be disgusting to every reasonable person." Another critic called the *Tractatus* "a book straight from hell" and its author the devil. The Catholic Church placed it on the Index of Forbidden Books.

So much censure could only work to heighten the reputation of its once-obscure author. A few years after Spinoza's death in 1677, his name was common currency among educated people; it was an established cipher for branding atheist, blasphemous, and dangerous ideas. Still, his works lived on, and his arguments became central to subsequent debate.

Spinoza's austerely rational humanism had brought philosophy into the here and now, as thinking in and about a physical world in which morality exists as a product of insight, a social contract between people balancing thoughts and desires. This epochal achievement met immense opposition, and it is certainly not the case that it marks a clear rational turn in European thinking. On the contrary. The religious debates saw an intensification of orthodox voices such as that of the immensely influential French court preacher Jacques-Bénigne Bossuet. For every book supporting rationalist or materialist positions (even if hidden behind various rhetorical smoke screens), there were several dedicated to refuting them, and to new biblical interpretations of history, politics, and nature, including a flood of pamphlets, sermons, and longer works dedicated to explaining the never-ending winters, dismal summers, and repeatedly ruined harvests as divine punishments and signs.

Persistent climatic pressure on natural resources had given rise to structural changes in human societies, which began to be more oriented toward markets and cities—more professionalized, and more dynamic. Trading cities such as Amsterdam, London, and Naples showed not only greater social mobility but also a market-based pragmatic tolerance and a broadening of intellectual horizons that was frequently associated with immigrants and minorities. This was the soil in which a largely secular, emphatically rational, and universalist philosophy could grow. Spinoza's life and fate are typical for this vast transitional period. He was the child of refugees, growing up in exile, introduced to a very different intellectual and social world; here he had to construct a new life, and he would enrich this life by approaching an old debate from a new perspective. This story was repeated countless times, in countless modulations, in cities all across Europe.

It may be surprising, in view of the climate crisis that was so often cast in a religious mold, that there were not more apocalyptic and millenarian cults and sects preaching the End of Days. Their comparative absence, however, is relatively easily explained. After a century of religious wars and sectarian violence, with a concomitant growth of both mysticism and militant sectarianism, Europeans had grown tired of extremist voices. The Inquisition had proven effective in silencing the most outspoken religious dissent, and with wars raging from the Netherlands via Central Europe to the Eastern Mediterranean, the faithful were tending to cling to a strong church that could possibly save their souls and also meet the more immediate need of feeding their bodies, if necessary.

There was, however, one messianic movement that reached a fever pitch, creating mass hysteria. It was an astonishing phenomenon that also illustrates the dangers minorities could experience when majorities came under too much pressure.

THE MESSIAH AND THE WHORE

The supposed Messiah who moved tens of thousands to believe fervently that the End of Days was nigh, and paradise just around the corner, was a tall young man whose life catapulted from inconspicuous beginnings to hysterical adulation, only to collapse in an inglorious end. His name was Sabbatai Zevi, and the adoring masses he attracted hailed primarily from Eastern Europe's Jewish communities.

Europe's Jews had no stake in the Christian sectarian conflicts of the sixteenth and seventeenth centuries, other than their own survival. As outsiders, they were always in danger of finding themselves between fronts, distrusted and even hated by both sides, sometimes accused of aligning with the enemy, sometimes the first to be attacked.

In Western Europe, many Jewish communities still carried the psychological scars of the brutal persecutions and forced exile during the fourteenth and fifteenth centuries on the Iberian Peninsula. Recent Sephardi arrivals, as well as Ashkenazi communities established in Europe for centuries, lived in cities rather than in the countryside, and they were relatively prosperous. In many cases, they were in active contact with their Gentile neighbors, as the example of Amsterdam has shown. Spinoza, an outcast from his own community, was not the only Jewish person in constant and friendly exchange with a wide range of freethinkers and scientists; Rembrandt knew and portrayed Jewish elders (which also means that they were relatively relaxed about Judaism's low opinion on creating graven images); an ex-Jesuit could be found teaching Jewish boys.

In Eastern Europe, where the largest Jewish communities were located, the situation was very different. Great numbers of Jews had arrived here from the West to escape persecution during the Middle Ages. These refugees had been assigned a Pale of Settlement in underpopulated lands ranging from the Baltic coast through modern Belarus, Ukraine, parts of Poland, and down to the Black Sea. Here they had lived more or less confined to their villages and shtetls, very much more segregated than the Jews of Western Europe. In most cases, they were forbidden to own land, to attend non-Jewish schools or universities, and to hold any kind of official position. In the villages, most survived on subsistence farming, or as craftsmen or laborers. In larger centers, Jews were able to support their own synagogues, together with the scholars who ensured that the religious heart of the community was kept alive. The synagogues also provided some elementary education, at least for boys, who were taught to read Hebrew and to recite prayers, but the social and often economic separation between Jews and Gentiles was strong, and poverty was endemic.

Jewish community leaders had been careful not to take sides in the religious wars and sectarian controversies of the post-Reformation era, but they had still paid a high price for them. They were obliged to pay special taxes to finance the wars, and to protect their own people from being robbed or even murdered, and every new wave of religious fanaticism, every famine or defeat on the battlefield was followed by a devastating tide of pogroms against them. During the 1648–1657 Cossack Rebellion under Bogdan Chmelnitski, some 100,000 Jews in the villages of what is now Ukraine were slaughtered by soldiers and local peasants. Some Christian Ukrainians—especially Catholic priests and Jesuits—also became victims of Cossack massacres, but the Jewish community bore the brunt of the rage and avarice of the Cossack mobs. Roughly a third of the Jews in this region were killed during these pogroms.

The widespread hatred of the Jews among Eastern European peasants was motivated not only by religious propaganda but also by economic factors that were intensified by climate change. Land-owners in the aristocratic Polish-Lithuanian Commonwealth, where the greatest concentration of Jews lived, were changing their agricultural production from subsistence farming to large-scale grain production for the international market. Harbors from Gdańsk to Königsberg and Riga were brimming with Dutch transport vessels loading up grain that would be shipped to Amsterdam and from there to all corners of Europe.

Like their English and Habsburg counterparts, the local aristocracy had found it considerably more lucrative to consolidate their lands by expelling tenants and sharecroppers and using the now-vacant land for their own commercial concerns. The peasants working the fields of Eastern Europe were mostly serfs, for whom the demands of the early Enlightenment were of no import whatsoever. After the Thirty Years' War, there were even several successful attempts to constrain further their already very limited rights. There was almost nothing here to impede increased farming along mercantilist principles.

In view of this situation, it is hardly surprising that the estate managers, whose task was to increase production and to fight old structures and practices in the name of ruthless efficiency, were especially despised among the peasants. Many of these estate managers were Jews, installed in these positions because they were literate and numerate, and also because they served as convenient lightning rods between the landowner and his irate and frequently desperate tenants and serfs.

Through the course of the seventeenth century, many peasants had consequently come to regard the Jews as the face of a new and oppressive order, and to blame them for the hunger of their families, as sacks of precious grain were sent abroad, frequently leaving too little to eat at home. Seeing Jews as estate managers and tax collectors, brandy producers and innkeepers (one of the few occupations they were allowed to have), most Gentiles found it too easy to associate them with their own exploitation. Jewish estate managers regularly failed to return home from their journeys, their bloodied bodies having been left in ditches after angry peasants had wreaked a brutal revenge.

For centuries, Jewish scholars had interpreted traumatic events as indicating the coming of the Messiah, who would finally arrive to rescue his persecuted people and lead them into a world of justice and happiness. At the height of the anti-Jewish pogroms and persecutions in Eastern Europe, just such an Anointed One suddenly appeared in the Mediterranean harbor town of Smyrna (today Izmir in Turkey).

Sabbatai Zevi (1626–1676) and his charismatic utterances caused one of the greatest mass hysterias of early modernity. The Hebrew name *Sabbatai*, meaning "the resting one," is derived from the planet

The Messiah? Sabbatai Zevi promised God's realm on earth, but died impoverished and forgotten in Albania.

governing the Sabbath (and the English Saturday), Saturn. According to mystical tradition, the rule of Saturn would usher in a time of peace and prosperity. Jewish mystics had long since integrated part of the astrological lore of antiquity into their own mystical tradition of the Kabbalah. Now the very name of the young man from Smyrna seemed to hold promise for those exhausted by the hardship and threats of their daily lives.

Zevi came from very humble origins—his father had owned a market stall, selling chickens. In 1648, the twenty-three-year-old began to speak, telling everyone willing to listen that he was the long-awaited Messiah. In order to prove his historic role, he broke a strict Jewish taboo, publicly pronouncing the name of God—without apparent divine retribution. To the few people who formed his early entourage, he read mystical texts and preached. The Jewish elders of Smyrna first kept him at arm's length, then banned him—first from the community and finally from the city itself. Zevi began to live the life of an itinerant prophet. In 1658, he appeared in the cosmopolitan city of Constantinople. Here another preacher began to support him, bolstering Zevi's claim by forging a document purporting to predict his coming, and stating that he would "cast down the great dragon" and would sit on God's throne.

This throne seemed quite out of reach, however, as Zevi was forced to keep on the move, always ahead of detractors and enemies, always on the lookout for support and new audiences. He traveled to Salonika, Alexandria, Athens, Smyrna, and Cairo before settling in Jerusalem in 1663, there creating a local sensation through his strict ascetic life and his curious exploits. He could be heard singing all night long—either Hebrew psalms or Spanish love songs, which, he claimed, symbolized his love of the Creator. His true fame, however, would only come when he returned to Cairo, where he found a well-to-do patron willing and able to promote his message to the world.

Suddenly materially secure, Zevi pursued his mission with renewed energy. During his stay in Egypt, he heard the tragic story of Sarah, a Jewish girl from Poland who had been orphaned during

the Chmelnitski pogroms, surviving only because she had been hidden by nuns, and then had fled. From Amsterdam, she had made her
way to Livorno, where she was forced to prostitute herself; during
this traumatic time, Sarah dreamed of one day becoming the bride
of the Messiah himself. Zevi, convinced that Divine Providence
had arranged matters this way, sent for the young woman because,
as he explained, it had been preordained that he was to marry a
fallen woman. He married Sarah in a lavish ceremony underwritten
by his patron, and then he returned to Jerusalem with her.

With his marriage, Zevi had fulfilled a prophecy and also built
an emotional bridge to the beleaguered Jews of Eastern Europe.
Surrounded by chaos, hatred, and uncertainty, they were immensely
drawn to the idea of a redeemer. The newlywed Messiah, who had
by now abandoned all thoughts of the ascetic life, traveled with his
supporters from Jerusalem to Aleppo, Gaza, and then to his home-
town of Smyrna, where he simply replaced the chief rabbi as leader
of the community. The great event would occur in 1666, he taught,
and it would be nothing less than the coming of a new era and the
homecoming of all Jews into the Holy Land. Nathan of Gaza, his
special adviser, even went one step farther: In 1666, he claimed, he
himself would place the sultan's turban on Zevi's head in Constantinople, where Zevi would become spiritual and temporal ruler.

A definite date for redemption created feverish anticipation
among the Jews in the diaspora. From North Africa and the Levant
to the shtetls of Eastern Europe and the great trading cities of the
West, many Jews were seized by messianic enthusiasm. Miraculous
stories about the ascetic Messiah and his beautiful young wife spread
throughout the Jewish world, and families thousands of miles from
Jerusalem sold their belongings in order to be ready for their final
journey home. The community elder of Venice, Moses Zacut (who
had attended school in Amsterdam with Spinoza), supported the cult
of the new Messiah; the community in Hamburg prayed for him three
times a week; in prayer books of the time, his portrait appears next to
that of King David; and in Poland the king found it necessary to for-
bid all public demonstrations in honor of Sabbatai Zevi.

The German scholar Heinrich Oldenburg, then living in London, regularly corresponded with Baruch de Spinoza. In one of his letters, he asks: "Everyone here speaks about the rumor of the return of the Israelites, who were dispersed for more than two thousand years, into their own country. Few believe it, but many wish for it. . . . If this news is confirmed, it would mean a great transformation."[14]

The prophet himself, however, was running into trouble. During another expedition to Constantinople, where his followers wanted to see him on the sultan's throne, he was understandably regarded with considerable distrust by the Ottoman authorities, who eventually arrested him. After several months in prison, he was granted an audience with Sultan Mehmed IV. When he still claimed to be the Messiah, the clever ruler confronted him with a simple choice. The Messiah was invulnerable, and the sultan proposed that Zevi allow an archer to aim for his chest. If the arrow did not penetrate him, the sultan and all his court would immediately convert to Judaism. But if the would-be savior refused to accept the challenge, he had to convert to Islam immediately.

Zevi and his faithful yielded to the greater force and became Muslims. Then, seeing his authority ebbing away, he tried to convince his followers that God had commanded his conversion, but fewer and fewer people were prepared to believe him. The movement that he had inspired collapsed, burying the hopes of tens of thousands of credulous contemporaries. Zevi himself was banished to Montenegro, to the little Albanian-speaking town of Dulcigno, where he died in 1676, destitute and forgotten.

THE FAIR ON THE ICE

Now, let us return to the immediate effects of the Little Ice Age in Europe. In 1684, the Continent witnessed the hardest winter in human memory. Lake Constance in southern Germany was covered in thick ice; in London, the architect and diarist John Evelyn wrote about the weather in the capital and beyond. His inventory shows the hard reality of living through Europe's long winter:

27 December, 1683. I went to visit Sir John Chardin. . . . It being in England this year one of the severest frosts that has happened of many years, he told me the cold in Persia was much greater, the ice of an incredible thickness. . . .

1 January, 1684. The weather continuing intolerably severe, streets of booths were set upon the Thames; the air was so very cold and thick, as of many years there had not been the like. The small-pox was very mortal.

6 January. The river quite frozen.

9 January. I went across the Thames on the ice, now become so thick as to bear not only streets of booths, in which they roasted meat, and had divers shops of wares, quite across as in a town, but coaches, carts, and horses passed over.

24 January. The frost continuing more and more severe, the Thames before London was still planted with booths in formal streets, all sorts of trades and shops furnished, and full of commodities, even to a printing-press, where the people and ladies took a fancy to have their names printed, and the day and year set down when printed on the Thames. . . . Coaches plied from Westminster to the Temple, and from several other stairs to and fro, as in the streets, sleds, sliding with skates, a bull-baiting, horse and coach-races, puppet-plays and interludes, cooks, tippling, and other lewd places, so that it seemed to be a bacchanalian triumph, or carnival on the water, whilst it was a severe judgment on the land, the trees not only splitting as if lightning-struck, but men and cattle perishing in divers places, and the very seas so locked up with ice, that no vessels could stir out or come in. The fowls, fish, and birds, and all our exotic plants and greens, universally perishing. Many parks of deer were destroyed, and all sorts of fuel so dear, that there were great contributions to preserve the poor alive. Nor was this severe weather much less intense in most parts of Europe, even as far as Spain and the most southern tracts. London, by reason of the excessive coldness of the air hindering the ascent of the smoke, was so filled with the fuliginous steam of the sea-coal, that hardly could one see across the streets, and this filing the lungs with its gross particles, exceedingly obstructed the breast, so as one could scarcely breathe. Here was no water to be had from the pipes and engines, nor could the brewers and divers other tradesmen work, and every moment was full of disastrous accidents.[15]

Evelyn was a thoughtful chronicler. While his entry for March 28 reads, "the weather began to be more mild and tolerable; but there was not the least appearance of any spring," he did not speculate whether the severe weather might be a sign of God's displeasure.

Frost Fair: During particularly severe winters the Thames became another part of London, a permanent fair on the ice.

He was hardly an intellectual revolutionary, but the thought did not seem to have occurred to him.

It was not only the sober John Evelyn who looked at nature without thinking about divine intervention. Two generations earlier, the North German pastor and astronomer David Fabricius had made a name for himself as an expert in observations of nature. He had spent his entire life among Friesian farmers, surrounded by meadows and sheep. In spite of the frequent and cumbersome travel necessitated by his pastoral duties, he still found time to observe the night skies, to correspond with scientists throughout Europe, and to redraw the maps of his local area. Fabricius also kept a detailed weather diary, which has been preserved. It details temperatures, cloud formations, and other salient information.

In 1611, his son Johannes returned from his studies in Leiden with one of the first telescopes in Germany. Fabricius then became one of the earliest Europeans to document variable stars, and he also made another discovery. With the help of the telescope, he could make out spots on the surface of the sun, which helped him to measure the length of the sun's revolution.

The demise of the pastor–scientist, however, was a very down-to-earth affair. According to witnesses, Fabricius had mentioned in a sermon that he knew who was responsible for a spate of recent goose thefts. On May 17, 1617, the pastor refused to leave his house. Like many of his scientist colleagues, he also dabbled in astrology, and he was convinced that his horoscope for this day was not propitious. Only toward the evening, when a different constellation had formed in the skies, did Fabricius feel safe enough to leave the house in order to visit a parishioner. En route, he was waylaid by a local peasant and battered to death with a peat shovel. The peasant, it seems, had felt threatened, because it was he who had stolen the geese.

The discovery of sunspots—made by Fabricius and independently also by other telescope-equipped stargazers throughout Europe—turned out to have special significance for the scientific understanding of the Little Ice Age, whose causes, even today, are not yet fully explained. During the second half of the seventeenth century, astronomers such as Giovanni Domenico Cassini, who worked first in Bologna and later at the Paris Observatory, had noticed an inexplicable phenomenon: The number of sunspots had greatly declined.

This lack of sunspots is known as the Maunder Minimum, named not after its original discoverers but after later interpreters. The period during which this phenomenon was observed—from 1645 to 1715—coincided with the greatest extended period of temperature reduction. There is still no generally accepted scientific explanation for this apparent correlation, but it appears that the frequency of sunspots is related to the intensity of sun activity, which would point to a change in solar activity during this time. This does not explain, however, why the Little Ice Age reached its first peak around 1570, and astronomers as well as climate historians continue to seek a comprehensive explanation for the historic great freeze.

THE FACE OF CHANGE

The Europe in which John Evelyn kept his diary was very different from that of 1570, when Wouter Jacobszoon had written his. By now, many people had emancipated themselves from the idea of a Creator controlling every daily detail of the natural world, and while most people who had heard Spinoza's name still shuddered when the arch-atheist was mentioned, it was common now, at least among educated people, to regard the natural world as a kind of giant clockwork whose mechanisms could be discovered and explained through careful observation.

A century had passed. Europe was still at war, harvests were still being ruined by severe weather, famines and epidemics still preyed on the Continent's people. Now, however, societies had begun to adapt to the changing circumstances, and the effects of these efforts would prove profound. For the first time in their history, European powers were becoming dominant in the world; their influence and curiosity, trade networks and wars, expeditions and ruthless colonial subjugation were penetrating ever farther around the globe.

These adaptations to climate change had occurred like other evolutionary processes: through trial and error. The initial religious and moral response to the climate had proven less effective than new agricultural techniques, new crops and property structures, new markets, harbors, roads, and canals. The most basic proposals for tackling the climate crisis came from gentlemen scholars we would today call botanists and agricultural engineers. Farming techniques that had changed little in the previous millennium were altered on the basis of experiment and observation, and their results were published and widely disseminated. Estate managers on large, consolidated farming estates were willing to experiment with new crops such as sweet corn and potatoes, with new patterns of land use, and with steel plows instead of wooden ones. Larger cattle herds enabled them to use oxen for tasks such as plowing and pulling carts, to have more manure for their fields, and to produce milk, cheese, and meat for markets in the towns.

Market-oriented farming increased greatly during the seventeenth century—from English and Habsburg lords expelling the peasants from the commons in order to raise sheep or produce grain on a large scale, to the Dutch farmers who reacted to climate change by giving up grain altogether and moving into cattle, vegetables, and flowers. The greatest area of agricultural production, however, was in Eastern Europe, where a sea of wheat stretched continuously from the famous black earth of Ukraine to the Baltic coast. The Little Ice Age made itself felt here with great severity, just as in the Western countries, but this region was less populated and its vast fields were tilled by unpaid serfs, allowing the landowners to sell at considerable profit to merchants from further West.

Agricultural methods were more efficient, and a significant rise in international trade made it possible to react to climate-related emergencies by moving grain and other goods in a Europe-wide market. Thus, cities in northern Italy could buy wheat and rye from Amsterdam when harvests in southern Italy had failed. The rise of international markets provided a mechanism for coping with crises, but it also caused a cascade of other, generally unforeseen changes—beginning with the increased importance of money in the economy.

This made life for the urban poor more precarious, but the workings of the new markets could also negatively affect those at the highest levels of power. Monarchs found themselves not only in ruinously expensive military confrontations with others but increasingly also in trade wars for raw materials, monopolies, and lucrative exports of manufactured goods. Capable rulers grasped the importance of professional administrators and other experts such as engineers and manufacturers, military leaders, philosophers, scientists, and artists. They attracted international talent by offering tax privileges and sanctuary to Huguenots, Jews, and other persecuted minorities with relevant skills or connections. France, for example, instituted a vast campaign of investment in roads, canals, and manufacturing, all while making a concerted effort to combat highwaymen and assert state control throughout its territory (including imposing French for the first time as the language in all areas of geographical France). Aided by growing professionalism and fueled by sophisticated fiscal systems, Europe's administrative states were beginning to emerge.

New ideas in turn spawned further ideas. Mercantilist authors had created the blueprint for a new vision of economic reality, in which trade was an aspect of war, a zero-sum game in which others had to lose for one of their own number to get rich; it was a principle that applied to individuals as well as to nations. This eternal battle of trading interests was fought initially inside Europe, but increasingly also overseas.

As more and more overseas territories were mapped and colonized, the race for raw materials transferred European rivalries to other continents. But, lacking the power or the reach to defend the vast international territories from rival powers, the Europeans instead agreed to carve up the globe into spheres of interest. The security gained for individual European powers by this new prac-

tice meant that even higher profits could be taken from the colonized territories.

Apart from the very wealthy, the citizens best able to take advantage of the new economic landscape were members of the broadening professional middle class, which allowed the most gifted and the
most determined to rise through the ranks: the artillery officer Descartes, the schoolmaster Bayle, the farmer's son Kircher, the weaver's son van den Enden, and the import-export merchant Spinoza,
among others we have already encountered.

A good many of the early developments were centered on
England (after 1707, Great Britain) and the Netherlands, the two
most liberal and most trade-oriented (and hence necessarily open)
countries in Europe. It was in these lands that the Enlightenment
practices first took shape, driven to some degree by intellectuals'
search for the truth but also by simmering social and cultural conflicts. By the second half of the seventeenth century, in most Western European countries, people working in the professions had
begun to tip the balance of power away from the aristocracy and
in their own favor. By now, they held most of the economic power
and a great deal of the cultural power, but still, in most places, very
little political power, which generally remained with the aristocracy
or the Church.

In this context, the arguments of thinkers we now might refer to
as "early Enlighteners" served a vital function. Ideas such as rationality, the rights of the individual, and equality before the law are as
old as philosophy itself, but they were never much heeded in rigidly
hierarchical societies based on the Bible and on birth precedence.
Now they were brought to the fore by a broad alliance of interests
allied in a fight—in the long run, a fight to the death—against the
old regime.

The ideas of the Enlightenment proved to be the perfect weapon
against the hermetic rule of the wellborn and the pious—a weapon
wielded by men and women who felt it was their right to control
their own fate and to have a say in how their communities oper-

ated. From the very beginning, philosophical debates about universal values were intertwined with more practical conflicts about local power and resources.

THE PRICE OF CHANGE

The complex transformation of Europe's societies that began during the most severe decades of the Little Ice Age was decisive in giving the global West its power, laying the foundations for four centuries of global dominance. The legacy of this ascendancy is still being discussed, but even during the seventeenth century, the conflict between philosophical ideals and economic interests at the heart of the rapid rise of European powers quickly became apparent. It is a conflict perfectly exemplified in the life and work of a provincial lawyer's son, the philosopher John Locke (1632–1704), another child from what we would now call a middle-class family.

The precocious John Locke had found a patron whose financial help saw the boy through school and then Oxford, where he studied philosophy and ancient languages, and also taught for a while, earning along the way a degree in medicine. At Oxford, the young man met some of the leading scientists of his day, including Robert Boyle and Thomas Sydenham; they turned his interests in a more empirical direction than most academic philosophers would have advocated at the time.

After his studies, Locke became private physician and secretary to Anthony Ashley Cooper, 1st Earl of Shaftesbury, an ambitious politician with progressive ideas. (His grandson, the 3rd earl, also named Anthony Ashley Cooper, would become one of the most significant liberal voices of the eighteenth century.) Taking part in the meetings of the Board of Trade and Plantations as His Lordship's trusted adviser, and relishing the practical challenges of political decisionmaking, Locke suddenly was close to real political power.

John Locke.

Human rights: John Locke made a delicate distinction between the rights of white Europeans and their black slaves.

He soon found himself enjoying the political games—so much so, in fact, that he was even drawn into a conspiracy against King Charles II. When the plot was discovered, he fled to Amsterdam. Here he remained for five years, probably coming in contact with the city's circles of freethinkers, then still very much in thrall to the ideas of Spinoza. Locke returned to England after the Glorious

Revolution of 1688, when William of Orange, invited to share the throne with the Protestant Queen Mary, sailed his fleet, quite peaceably, up the Thames to London.

By now, Locke had already published several philosophical works. His writings also dealt with the foundation of and justification for political power. In the second of his *Two Treatises of Government*, he analyzes the fundamental properties of humankind. Being God's creatures, he writes, people are necessarily equally free and equally valuable by nature: "men being . . . by nature, all free, equal and independent, no one can be put out of his estate, and subjected to the political power of another, without his consent."[16]

These words reflect the experience of a political refugee, and also the influence of the intellectual atmosphere during his time of exile, as do the following:

> The state of nature has a law of nature to govern it, which obliges every one: and reason, which is that law, teaches all mankind, who will but consult it, that being all equal and independent, no one ought to harm another in his life, health, liberty, or possessions: . . . and being furnished with like faculties, sharing all in one community of nature, there cannot be supposed any such subordination among us, that may authorize us to destroy one another, as if we were made for one another's uses, as the inferior ranks of creatures are for our's.[17]

The laws of a just society are the laws of nature, even if cooperation requires that people leave the state of nature and its liberties and enter into a contractual relationship with one another.

It would be hard to overstate the trailblazing daring of these ideas, whose echoes are found throughout the Enlightenment and in the wording of the US Constitution, but their author was coolly pragmatic. While in his writings he argued that every person is equal and free by nature and has an inalienable right to life, liberty, health, and property, the administrative duties of his new employment in London also made him an investor in and administrator of plantations in Carolina, and as such an important actor in colonial slavery. In 1669, he even wrote, together with Shaftesbury, a constitution for the province of Carolina, which stated in article 110: "Every freeman of Carolina shall have absolute power and authority over his negro slaves, of what opinion or religion soever."

Locke's apparent cynicism does not imply that he was not convinced of the natural freedom and equality of all people—as long as they did not touch on the political and business interests of the Crown. As a philosopher, he was a man of principle; as a royal administrator, he kept his eyes firmly on the bottom line.

To the great majority of his readers, Locke was a thinker, not a politically compromised doer. His philosophical arguments were widely read and discussed, and they served an important function in social and political debates. In the premodern world, it was always the group, the many, who were right. The village, tradition, the Church, one's social station, the guild, the power of ancestral custom—each decided what was acceptable and what was done, setting its stamp on all aspects of daily life.

Confessional disputes, the Inquisition, the expulsion of those of a different faith, and the oppression of minorities in the name of the One True Faith had made continuing peaceful coexistence almost impossible, and business much harder to carry on. Trading centers such as London and Amsterdam could only flourish (as Voltaire would later write home enthusiastically) because Catholics could trade with Jews and Huguenots, and Lutherans with Muslims, and because they all knew that they could by and large trust one another—and the laws protecting them all. Europe's urban centers needed the concepts of coexistence and tolerance if they were to grow into successful larger and more complex societies.

Here the ideas of Spinoza and Locke played a decisive role. Their concept of universal human rights turned the conventional social order on its head by strengthening the rights of individuals against collectives. Nobody had the right to rob an individual of his or her freedom and dignity; no single religious truth could dominate all others. Only individuals, endowed by nature with equal and inalienable rights and freedoms, could meet as individuals and not as representatives of social, ethnic, or religious collectives, especially since these were frequently antagonistic in ways that the individuals simply did not share.

TAPISSIER DU ROI

The hypocrisy of an English philosopher who distinguished sharply between his theoretical writings and his practical concerns has damaged Locke's posthumous reputation as a father of universal human rights, but it was typical of the thinking of European elites, and later of that of the bourgeoisie, whose wealth so often rested on slave labor and the structures supported by it. Locke's personal moral failure did little to disqualify him from becoming a moral exemplar—indeed, his own glaring inconsistency mirrors that of later generations.

This inconsistency, arguably in fact hypocrisy, was a popular target for satire as well as a cause of social discontent; the official image of power was so different from the reality behind it that eventually the authorities themselves initiated a major new public relations effort. They began to commission works of art representing them in a different light: as heroes of the ancient world, and even as gods. Thus was born a new culture of aristocratic display, in the art of the Baroque.

Before the seventeenth century, rulers had frequently been portrayed relatively realistically, in stylized but recognizable surroundings. Now Europe's monarchs—generally grown men, and often enough with double chins and distinctly unheroic bellies— appeared in paintings or as statues half-naked or draped in togas, with armor recalling the great generals of ancient Rome. Or they stood surrounded by angels, muses, and allegories of justice, wisdom, mercy, and virtue. A monarch's claim to the throne was now to be fully dramatized. Once understood as being implicitly present in the ruler, his virtues now apparently needed to be seen to

be believed. The Sun King, Louis XIV of France, was one of the
most lavish proponents of this fashion. During the early years of
his reign, he frequently appeared on stage, once resplendently cos-
tumed as Apollo.

Especially in the Protestant North, large sections of the bourgeoisie found this kind of wastefulness instinctively distasteful. The
northern pictorial universe was also allegorical, but it was a great
deal more sober, as countless winter landscapes and still lifes attest.

Bourgeois art, as we have seen, was more concerned with mor-
alistic themes of earthly transience than with grafting the legends
of antiquity onto the faces of modern sovereigns. But middle-class
moralism had its own inconsistencies, as the literature of the day
delightfully reveals. If dwelling piously on the next life had its recognized place in bourgeois life, so did determined social climbing.
Wealthier bourgeois tried to imitate the aristocracy in its manners
or speech, or even to rise into it by marriage or the purchase of
titles and estates. Bourgeois ambition, and the anxiety attendant on
it, became a favored trope of satirists. The French actor and playwright Molière, in particular, took the greatest pleasure in skewering the big pretensions of the rising classes, along with their small
dishonesties.

Oversexed clerics, old misers, teenagers with ideas above their
station, mendacious servants, and hypocrites of all stripes form the
casts of his plays, which were all the more irresistible to contemporary theatergoers because they recognized these characters from
their own lives. Molière knew exactly what he was writing about.
His own father had been a classic social climber, an interior designer
working for the nouveaux riches and the aristocracy, an occupation
that finally allowed him to call himself *Tapissier du roi.* The social
anxiety that runs through Molière's plays was familiar to him from
his own family.

At his court of Versailles, the Sun King was in an ideal position
to observe the dance of vanity and hypocrisy among those seeking advancement, looking for sexual adventures, offering bribes, or
hunting sinecures. He made Molière director of entertainment, and

Sun King: Louis XIV was unsurpassed in the art
of staging his own glory, here as Apollo.

in 1664 his comedy *Tartuffe* was performed during an elaborate garden party that lasted three days.

The play, in which a man pretending to be deeply pious almost ruins an entire family, caused such a scandal at court that Louis, who had enjoyed the performance, was forced to ban it from being performed in public. For years to come, the playwright would fight hopelessly to get his play performed. The gap between the stage and real life had become too small, and that between official and lived morality too wide.

THE PUBLIC SPHERE AND THE VICES OF BEES

The theater was on the road as well, with traveling companies playing in market squares, allowing countless ordinary people to reencounter familiar storylines and meet new ideas. In this, the theater resembles other aspects of the developing public sphere: Never before had there been such a flood of philosophical and political tracts, newspapers, almanacs, sentimental novels, political pamphlets, and pornographic satire. Since paper was cheap and easy to transport, ideas could travel quickly, with geographical distance overcome by shared languages—at first mainly Latin, and later French and English. This formed a new kind of community, a borderless "Republic of Letters" that was, in theory at least, a home to all who could read, or who had someone to read to them.

It was largely through this Republic of Letters, rather than in the conservative universities, that the materialist thinkers of the seventeenth century disseminated their radical ideas. Philosophical materialism has a tradition reaching back to early classical antiquity in Europe, and related traditions also exist in Islamic and Chinese thought. Earlier thinkers, though, were often subject to censorship and could publish their ideas only in the most diplomatic and roundabout ways. More important, they had no opportunities to teach or even discuss their ideas beyond their own radical circles. Religious authorities everywhere, and the rulers to whom they gave support, regarded these writings as dangerous, since they denied any divine origin for the legitimacy of monarchies and other worldly powers.

The centuries-long relative isolation of materialist thinkers meant that, especially in less populated places, they often had to find their own intellectual way to these ideas. In Europe, most early modern materialists found inspiration from a small number of Greek and Roman texts, especially Lucretius. This changed, however, with the emergence of the Republic of Letters in the seventeenth century. Official books and newspapers were relatively easy to censor and control, but printing presses were not, and these were in service day and night, producing broadsheets and flyers, pamphlets and polemics. For the first time, the printed word had escaped the control of the authorities, temporal and spiritual.

The increased availability of cheap printed materials was accompanied by a dramatic rise in literacy, at least in some countries. In Northern and Western Europe, three times as many children, most of them boys, were now sent to school than in the previous century. Some of these children later went into the professions, into the army, into trade, or into the growing branches of administration; others became skilled workers in areas where literacy offered significant advantages, such as printing, the navy, or copying and secretarial work. Worries about the effects of sentimental novels on domestic servants in the Netherlands illustrate how widespread reading had become, and how receptive to new ideas people were judged to be.

This period also witnessed an intensified debate among voices from different countries. Authors who had once worked in seclusion now had access to a wider public. Even if censorship frequently imposed anonymity, the ideas themselves had now seen the light of day, and ideas are gloriously promiscuous. Entire intellectual traditions began to grow, branching out in new directions. Spinoza would have been unthinkable without Descartes and the Amsterdam freethinkers, without Jewish thought and the ongoing formulation of scientific principles, and his work would be carried on, seminal material for generations of thinkers. Spinoza's ideas could be attacked, but they could not be unthought.

A flourishing market in anonymous libels, serious reports, edifying books, new translations, popular fiction, poetry, and other literary genres brought new kinds of actors onto the scene, people who lived by their wits and by their pens. Journalists, anonymous hacks, philosophers, pornographers, and pious preachers all relied on commissions—a few guilders or shillings here and there—to keep themselves in bread and wine, and they tended to congregate in cafés, on public plazas, and in the more discreet setting of private salons. The Republic of Letters had found its citizens, and they were a cantankerous lot.

> Happy is the Man that has no other Acquaintance with Booksellers, than what is Contracted by Reading the News in their Shops, and perhaps now and then Buying a Book of them; but he, that is so unfortunate, as to have Business with them about Translating, Printing, or Publishing any Thing to the World, has a Miserable Time of it, and ought to be endowed with the Patience of *Job*.[18]

Bernard Mandeville (1670–1733), the author of these lines, made a specialty of being argumentative, provocative, and alarmingly straightforward—about booksellers (who also were publishers at the time) and also about matters ranging from legalizing prostitution to hysteria and the death penalty.

Mandeville was born in Rotterdam and studied medicine in Leiden before choosing to settle in London. Next to his medical practice, he made his name as the author of several remarkably frank and pragmatic pamphlets. In *The Virgin Unmask'd*, he described marriage as a sexual prison for women, and in *A Modest Defence of Publick Stews*, he suggested that all "stews" (brothels) be legalized and medically controlled.

These views caused several scandals in London society, a fact that appears to have delighted the Dutch physician, who had imported his country's unideological perspective to his new English home. He was an elegant provocateur who could write an entire preface to

one of his books about how much he hated writing prefaces, who
wrote about the clitoris as the center of female lust, who refused to
see lust as a moral question at all, and who poured many of his controversial thoughts into doggerel. Mandeville's most famous work is
also written in verse. It begins, harmlessly enough, with the image
of a beehive, which bears a striking resemblance to a seventeenth-
century mercantilist state:

> A Spacious Hive well stockt with Bees,
> That liv'd in Luxury and Ease;
> And yet as fam'd for Laws and Arms,
> As yielding large and early Swarms;
> Was counted the great Nursery
> Of Sciences and Industry.

The secret behind this humming idyll is simple: The bees work hard
in order to profit from the cupidity, the stupidity, and the vanity
of others. The beehive's wealth is a direct result of the efforts of
"Sharpers, Parasites, Pimps, Players, Pick-pockets, Coiners, Quacks,
South-sayers." The hive thrives on dishonesty, flattery, vanity,
and deceit.

All professions practiced in the beehive are profoundly corrupt:
Lawyers prolong cases unnecessarily by splitting hairs; doctors are
incompetent toadies who demand horrendous fees and are at the
same time killing their patients with their potions; priests are stupid,
lecherous, and greedy and still remain poor; soldiers are either brute
cannon fodder or arrogant cowards; and judges are busy handing
down severe sentences to the

> . . . Desp'rate and the Poor;
> That, urg'd by meer Necessity,
> Were ty'd up to the wretched Tree
> For Crimes, which not deserv'd that Fate,
> But to secure the Rich and Great.

But the result of all this knavery was surprising:

> Thus every Part was full of Vice,
> Yet the whole Mass a Paradise;
> Flatter'd in Peace, and fear'd in Wars,
> They were th' Esteem of Foreigners,
> And lavish of their Wealth and Lives,
> The Balance of all other Hives.
> Such were the Blessings of that State;
> Their Crimes conspir'd to make them Great.

This paradisiacal and paradoxical state is rudely interrupted when one bee decides to end the rule of vice and convert its fellow bees to a life of virtue and honesty. As if by a miracle, this works almost instantaneously. The bees become virtuous, modest, and truthful. They develop a conscience and confess their crimes to one another.

But the consequences are not quite what one might expect. Under the influence of this new apostle of morality, fraud and theft decline, prices go down, great gentlemen release their vast retinues of servants, and sell their pompous coaches and the gold-tressed liveries of their footmen. Country estates lose their value and all bees now dress simply and modestly and stop buying luxurious and exotic goods. Overseas trade withers away, and even at home the merchant bees soon get into trouble; craftsmen and tradesmen give up, the state loses taxes and influence, unemployment leads to hunger and misery among the worker bees. In the end, the last survivors move into a hollow tree, bereft of the blossoming society they once inhabited, yet convinced of their moral superiority.

For the author, who is obviously enjoying his loving description of sinful insects, the Moral is clear:

> Fools only strive
> To make a Great and Honest Hive
> T' enjoy the World's Conveniencies,
> Be fam'd in War, yet live in Ease,

Without great Vices, is a vain
Eutopia seated in the Brain.
Fraud, Luxury and Pride must live. . . .

In a time when public discussion was still dominated by a strongly religious rhetoric of Christian virtue, the sacredness of life according to the Bible, and the fight against vice, Mandeville had wittily upended the moral scale: The beehive can only thrive if the animals follow their lowest instincts, and because they follow them. Virtue would only wreck the great bonfire that keeps burning merrily, only for as long as there is vanity to fuel it.

In his robust pragmatism, Mandeville anticipated a social vision that would become dominant long after his death, while other thinkers argued, often inspired by him, that egotism and self-interest, not virtue, are the driving forces of societies, and the interplay of individual self-interests can and will result in achieving the common good. Mandeville even went further, describing Christian virtues as stupid and destructive, the instruments of a powerful elite that reinforced its power with the aid of churchmen who were either cynical or brainless. Together, they prevented society from flourishing. A message of God-given truth and law, which had been the only one that was visible and acceptable, had become only one possible message of many. Moreover, as Mandeville dared to insinuate, the religious message had become rather annoying.

THE FLOATING REVEREND

The Little Ice Age was to last for roughly a hundred years more. Although it was a global phenomenon, I have focused almost exclusively on Europe, because it offers the most varied and most finely grained information about the interrelation between climate change and cultural developments. The peak of the climate episode we know as the Little Ice Age coincides with massive changes in European societies. To some extent at least, improved agricultural techniques, stronger and more international markets, and an increasingly globalized system of economic domination (of growth based on exploitation) allowed Europeans to develop more successful responses to climate change and to the hardships it inflicted. These responses were answered in their turn by transformations in every aspect of culture and society.

There was another wave of cold winters and summers during the early eighteenth century, but by that point, the social effects of such events tended to be less severe for Europeans. There were, however, some notable exceptions, such as the years preceding 1789 in France, which left the population hungry and increasingly embittered. Still, during the eighteenth century in Western Europe at least, there were fewer famines than there had been during the previous two centuries.

Comparative historical climate data suggest that the cooling effect was particularly strong in Europe, the most northerly land mass inhabited by great numbers of people, who were especially vulnerable to frigid winters and food shortages. The intensity of the temperature change meant that finding coping mechanisms was an existential challenge. Perhaps the sense of urgency expressed by so many European voices created a greater willingness to enter the unknown—to experiment, to debate, and to risk change.

Interestingly, during the same period, severe winters, devastating famines, and wars and civil wars in Russia, China, and the Ottoman Empire did not result in comparable changes in agriculture, the economy, and culture. And so, as the story goes, the West won global dominance through qualities and ideas intrinsic to European culture.

But historical facts are never quite as smooth as grand narratives require them to be. The Little Ice Age is also a story of accidental adaptation, of coincidental circumstances allowing great transformations to take place. In Russia, for instance, the bloody civil wars in the west of the empire were also a power struggle between Westernizers and traditionalists. Nor would an overseas colonial empire have been an obvious goal for the Russians: Their hinterlands were already vast, and, crucially, they had no easy access to the open sea.

In China, failed harvests and the cruel famines that followed contributed to the collapse of the Ming dynasty in 1644, but foreign trade still continued and indeed increased, and successive emperors were able to build up and support a remarkably effective administrative state. After the Manchu emperors seized power in 1644, the country was further disrupted by internal resistance to these foreign rulers, yet both economic and intellectual exchanges—including with European merchants and Jesuit missionaries—continued unabated. As far as geographical expansion was concerned, China had quite enough territory to develop and defend as it was, and neighboring countries such as Korea could be made into vassals more readily than others across the oceans.

The Ottoman Empire suffered greatly during the abnormally cold winters, severe enough even to freeze the Bosphorus Strait. Harvests failed, and famines, epidemics, and rebellions ensued, but the empire was already in trouble in other ways. International trade had been moving away from the Eastern Mediterranean and into the Atlantic and the Pacific, severing not only the historic relationship between Christian Europe and Islamic civilizations but also relegating the once-pivotal Ottoman lands to the periphery of world trade. Of necessity, this shrank the sultan's ambitions. When the Ottoman Siege of Vienna in 1683 ended in defeat, the sultanate was forced to abandon its grand ambition of invading Europe. There simply were not enough resources, and not enough loyal troops, to contemplate so vast an undertaking.

The Ottoman Empire was renegotiating its relationships with neighbors on its western, eastern, and northern borders while having to contend with a major food crisis and contain the volatility of its own outlying provinces and even allies. Like Russia, the Sublime Porte also had no direct and easy access to the oceans. There were no doubt cultural factors facilitating or hindering the development of new ideas and technological adaptations, but geography, for instance, is likely to have played an equally important role in deciding the destinies of these very different regions.

With easy access to the open seas, European rulers, traders, and adventurers had clear advantages over their potential competitors. A progressive intellectual culture may have hastened Europe's path to global trade dominance, but even if Russia, China, and the Ottoman Empire had been steeped in that same culture, geographical and other disadvantages may have prevented a similar rise on their part.

But a positive attitude toward continuing intellectual experiment was vital. As in Clusius's exotic botanical garden, curious specimens would first be planted in a few simple rows: Could they be cultivated, could they be eaten, smoked, or used for fibers, medicines, or intoxicating substances? Most would not survive their first winter, but some would thrive, and a few would become the staples of European gardens and fields. Success was often accidental—perhaps dependent on a chance observation, or on the long-term dedication of one bold or determined experimenter—or indeed on both.

Then as now, scientists on the outer limits of their disciplines postulated theories that seemed extreme. Then as now, these theories could not explain everything that could be observed. But the resources that states, churches, and private people were investing in education created a public sphere where new ideas could be discussed, and practical questions could be addressed alongside, and often emerging from, the new philosophical and political thinking. For the first time, these broad-based efforts to educate reached a significant number of people from very humble backgrounds. The historical consequences would be tremendous.

A great deal of this transformation was prompted, whether consciously or not, by responses to the Little Ice Age. Vintners moved their grapevines southward about 250 miles (400 kilometers) to find the sunshine necessary for the fruit to ripen. Agriculture diversified to include more robust plants such as potatoes and maize, and new techniques of cultivation increased yields of grain by up to a third. Producing for the market changed rural economies, and the grain trade helped to address food supply problems; money and expertise concentrated in the cities. And overseas trade, with populations willing and unwilling, made Europe ever richer.

The motivation for the smaller-scale changes was often simply physical survival, though sometimes, as in the case of poorer Protestants being taught to read the Bible, it was ideological. For the large-scale transformations, the motivations were money and retention of power. The surge in Catholic education, for instance, had little to do with forming model citizens or even saving souls; rather, the Catholic Church needed a larger cadre of its own to counter the Protestants' theological and biblical arguments.

Changes in military tactics, administration, and trading habits meant that those in power needed at least part of their population to be literate and numerate, to avoid falling behind in their fierce mercantilist competition with other ambitious powers. And, as so often occurs, one change triggers the next: The new layer of educated specialists and professionals soon realized that their own social interests lay not in serving as managers for the old order but in creating a new order, one in which they themselves would take charge.

Societies across Europe adapted to these changes at varying speeds, and with varying success. In Russia, in parts of the Ottoman Empire, and also in conservative Catholic Spain, reforms were crushed, as elites largely (despite the modernizing efforts of Tsar Peter the Great in Russia) decided to put their efforts and resources into preserving the status quo.

Reactionary Spain and the progressive Netherlands mark the extremes of the transformative changes taking place throughout Europe. Other regions reacted differently, with correspondingly different outcomes. In the central German-speaking regions, the Thirty Years' War caused dramatic population losses, and fragmented power structures made wide-ranging social change impossible. But war was not always an impediment to change: Riven by military and religious violence, France nevertheless created a strong, mercantilist state, as did England—not only during the reign of Elizabeth I but also during the English Civil War in the mid-seventeenth century and the Glorious Revolution forty years later.

Whether or not this landscape of economic and social transformation constituted progress, in the sense of improving material conditions, is an important question. It depends very much on who in society was experiencing the changes. The expansion of strong markets and the beginnings of an administrative state brought social emancipation for the more educated city dwellers, and initially it consolidated the power of aristocratic rulers. Eventually, the advancing middle classes would claim their former rulers' social and then political power. History, generally written from the perspective of its winners, too often tells the story of the rise of the most fortunate as though it had been the story of the rise of all.

For the rural poor, for day laborers, for farmers fleeing wars and religious persecution, and for those living in Europe's expanding network of colonies and slave plantations, the rise of newly efficient markets, administrations, and military tactics was a near-unmitigated disaster. The enclosure of commons and the consolidation of landed estates in Europe, increased taxation, inflation, deadly epidemics spreading in the hungry countryside or through cramped urban dwellings, to say nothing of the violence routinely unleashed against populations overseas—all this meant that, for most people, daily life became harsher.

While cities were swept along by the dynamism of new ideas and expanding trade, agricultural communities remained, on the whole, profoundly conservative. Farming techniques (and, in some areas, patterns of ownership) had changed, but minds were just as likely to have grown more fearful, more welded to tradition, faith, and custom. Compounding the difficulty, even in times of peace and comparative plenty, the cities offered the chance of a broader occupational and social life to the young and ambitious, and to those more open to new ideas.

Areas lagging in the adaptive process, keeping to their old power structures and insisting on maintaining traditional ways, could pay a high price for their reluctance to change. We see this, for instance, in the introduction of the South American potato to Europe. France was one of the last countries to accept the new crop, though it was hardier than wheat and less likely to fail in colder seasons. Rural populations distrusted it as something new and foreign. The landowners—mostly aristocrats obliged to remain at court as Louis XIV worked to centralize his power—had little interest in farming, and they were on the whole not present to oversee or even suggest the introduction of a new crop. The contrast with the farming-obsessed great lords of England is instructive. During the 1780s, extreme hardship after a series of failed harvests in France contributed enormously to the resentment and desperation that fed the Revolution of 1789.

><

THE FAMOUS YEAR without a summer was 1816, which many climate historians now regard as the final manifestation of the Little Ice Age. Severe frosts in May of that year damaged crops in Europe and destroyed most crops in North America. The English author Mary Shelley spent these sunless, rainy summer months at Lake Geneva with her husband, the poet Percy Bysshe Shelley. The cold forced them to stay indoors, and Mary found the atmosphere so oppressive that she set about writing a horror story about human hubris and the revenge of nature. She titled her book *Frankenstein—or, The Modern Prometheus.*

Although the dismal conditions of 1816 may have been due to the last manifestations of a general global cooling, the immediate cause was a vast cloud of ash hurled skyward by the eruption of Mount Tambora in Indonesia. In April 1815, a gigantic detonation had blown the entire top off the mountain, reducing its height by nearly five thousand feet (fifteen hundred meters). The eruption could be heard on Sumatra, 1,200 miles (nearly 2,000 kilometers) away, and islands in the region were hit by a tsunami thirteen feet (four meters) high. Millions of cubic feet of sulfurous ash reached the atmosphere, where they were dispersed by high winds, reflecting back much of the sunlight traveling toward Earth.

With the last effects of the Little Ice Age, some of its cultural manifestations also waned. This book began with a winter landscape, at the time a new artistic genre, by the Dutch painter Hendrick Avercamp. One of the last great winter landscapes was painted in the 1790s by the Scottish artist Henry Raeburn, portraying his friend Reverend Robert Walker, a passionate ice-skater. Walker had spent part of his childhood in Rotterdam, and it is likely that he had encountered skating for pleasure in the Netherlands and had perfected his skills there.

In the painting, the minister is wearing long Dutch skates tied over his shoes. He is gliding across the ice, seemingly without touching it. His body is tilted forward, out of balance—perhaps leaning toward the future. The ice beneath his skates is scratched with the traces of impressive pirouettes. The Scottish winter light softens the hardness of the ice, and the edges of the lake appear to blur into the surrounding hills. The skater, however, does not even seem to notice the landscape around him. Looking far ahead, he is hurtling out of the picture, caught in the moment, absorbed in thought, full of confidence in his ability to negotiate the vast plain in front of him, held upright and racing forward by the mathematical certainty of the laws of nature.

He is alone.

EPILOGUE
Supplement to The Fable of the Bees

SONGBIRDS, WOOD LICE, AND CORALS

The world as a beehive whose individual members are serving the greater good by lying and cheating—the acuity of Bernard Mandeville's *Fable of the Bees* lies not in its rhyming insults but in its moral intuition. We are animals, the author suggests; we do not live according to some abstract moral rule, but rather according to our natural instincts and desires. It is senseless to try to transform human beings into something they are not, and can never be.

To the eighteenth-century observer, the result was paradoxically beneficial as exploitation and cheating created wealth, but, equally important, the footprint of eighteenth-century economies was not large enough to cause irreversible changes to the Earth's ecosystem. In Mandeville's time, it was correct that maximum growth carried huge benefits for the swarm, and that selfish motives tended to accelerate this growth.

This detached, "apiological" perspective on history allows us to look at entire societies and their transformations without even posing one very important question: Did these changes constitute progress? Were the ideas growing out of them—capitalist markets, human rights, the Enlightenment—good or bad? What exactly did they change, and whom did they help?

Like bees, mosses, songbirds, wood lice, and corals, human

populations are part of the ecosystem and must adapt to local conditions. The only difference between us and other animals is that
Homo sapiens has a stronger ability to accelerate evolution, not only
by relying on genetic adaptation but also by manipulating ideas and
transforming cultural practices. Human nature may not change, but
human behavior can and does: Within a few generations, we can
learn to act differently, to survive in new surroundings, to plan and
make strategies for a future we can anticipate and try to comprehend.

><

IF WE (with hats raised to Mandeville) look at Europe between 1570
and 1680 as a giant beehive in which different swarms existed next to
one another—always trying to build their own honeycombs higher
than the neighbors', always threatening and skirmishing, battling
to survive—we can imagine what a gigantic collective effort these
countless individuals made.

Before the "big freeze," these swarms inhabited a small corner
of the meadow, but when food became scarce and unreliable, they
spread farther, colonized distant areas, used different kinds of food,
kept tighter control on their worker bees, and acquired more worker
bees elsewhere to become slaves.

This became a game of collecting treasure for the queen, not
only to quench her insatiable appetite but also to lay the foundations
for the future greatness of the hive. Whereas initially this may have
been a strictly local affair, now more effort goes into controlling and
enslaving distant hives in order to harvest their resources and, effectively, to live off their work. Some of the bees have become traders,
some are grim soldiers guarding the precious stocks and those who
carry them, others have specialized in counting, calculating, and
administering the hoard. Distant hives are brutally enslaved by this
regime as the conquering swarm grows in strength and numbers. Its
culture flourishes. It produces great works of art, revelatory scientific discoveries, and a culture of reflection that begins to question
its own privilege. Therein lies its moral greatness.

※

WE CANNOT WATCH these little creatures think, but we can never-theless admire their adaptive success. Without any great plan, with-out even understanding that they had entered a time of systemic, global climate change, Europe's societies found ways of coping with the new circumstances. They created early capitalism, a new eco-nomic order. An educated middle class rose in the cities and began to change society according to its own interests; cultural and politi-cal practices changed, institutions were founded, markets strength-ened, new intellectual horizons opened up. Within two generations, a formerly feudal continent of landowners and peasants was on its way to becoming a region of market power, urban energy, and Enlightened political and moral ideas.

The road from an agricultural crisis to capitalist markets was not a straight one; indeed, it was less a road and more a maze of paths converging on what proved to work best, at least for those at the top. This development had surprising side effects. As the natural order appeared to have become unhinged, and unseasonal weather upset traditional expectations, natural phenomena were observed with more attention, and on their own terms. In the minds of those try-ing to understand it, nature had emancipated itself from Creation. For centuries, the divine order had been used to justify inflexible social hierarchies and to limit the scope of legitimate intellectual inquiry. Thinkers such as Spinoza, however, used these same argu-ments to demand freedom, equality, tolerance, and the identifica-tion of God with nature, making science and philosophy central to civilized life.

As Karl Polanyi pointed out, Europe's relationship with money also changed significantly. Whereas previously most rural popula-tions had only infrequent contact with money and markets, chang-ing patterns of landownership meant that money, or the lack thereof, was beginning to have a greater role in people's lives.

For city dwellers, this new reality significantly changed their outlook on life. Markets encourage pragmatic tolerance in religious and other matters, while demanding a regime of equality before the law, of rules and regulations being enforced. While they tend to treat people as individuals with choices, they also dramatically limit the choices of those unable to participate. Medieval societies recognized that the poor had their place in a Christian society. For market societies, however, the only use of those too poor to consume was to depress wages, thus creating greater profits for the rich.

Within these new societies, those with sufficient ability, charisma, or ruthlessness could rise much faster and much higher than their fathers or grandfathers had done, but this mechanism only worked in conjunction with education or capital. Those who had neither land nor marketable special skills soon found themselves part of a vast reservoir of poor laborers who could be hired and fired, employed in manufacturing, expelled and resettled, used as chattel in diplomatic negotiations, or pressed into the vital navies of the mercantilist powers.

Destitute and very often close to starvation, these powerless members of society had no choice but to do whatever work they could find—no matter how hard, how dangerous, or how badly paid. Coordinated rebellion was rare, and always ruthlessly suppressed.

Paradoxically, mercantilist writers were constantly preoccupied with the problem of the poor—not because they wanted to end poverty but because they knew they depended on it for cheap labor. But even the poor cannot be productive if they are starving. They must at least earn enough to feed themselves and raise the next poor generation. Poverty was seen almost exclusively as a management issue, not a question of social justice.

The market economy, habits of empirical observation and scientific thinking, and—gradually—ideas of equality, tolerance, and human rights changed European societies and put new elites in positions of power. Access to the oceans allowed for colonial expansion, and a strong system of economic exploitation at home and abroad produced economic growth and created a layer of social and financial wealth out of which modern states and institutions would grow.

During the sixteenth and seventeenth centuries, the idea of human equality, so widely accepted today, not only was novel but also was regarded as politically subversive and even blasphemous to religion. A natural law protecting the intrinsic, equal, and inalienable rights of all individuals—of their conscience as well as their property—was politically suspect and religiously problematic; it was rightly perceived to undermine the dominant social order.

The ideas blossoming during this time (ideas we now tend to classify as "Enlightenment" ideas) were also used as weapons in social and ideological conflicts. Tolerance and freedom of conscience were important tools in the fight against the power of the Church; political freedom and individualism allowed literate city dwellers to set limits on aristocratic power and to establish a power base of their own.

As always, these ideas were also used to justify power as well as oppression. Colonial rule could and was argued for by reference to Christianity as a holy mission, or as a cultural mission (*la mission civilisatrice*) to save religious heathens and cultural savages from themselves. As we have seen in the case of John Locke, noble rhetoric about human rights and universal equality was not allowed to get in the way of business; Enlightenment ideals were quickly embroiled in dirty, messy compromises.

It would not be long before those at the bottom of this new social and colonial pyramid would begin to quote back the principles of the Enlightenment to their oppressors, and even to try to resist their oppression in the name of these ideas. Whenever this happened among rural populations, or later, the urban working poor in Europe, or among colonized peoples, the answer was almost always the same—swift and bloody.

The dream of universal human rights had once been dreamt mainly by the middle classes against the aristocracy. It proved a powerful idea, but its sword was double-edged. The intellectual success of the European (now the Western) model rests on its universalist ideas. Its economic success, however, rests on suspending these ideas for the larger part of humanity. As economic growth based on exploitation became the magic formula of Western dominance, more and more energy and sophistry had to be invested in appearing to reconcile it with the noble ideas of the Enlightenment.

FREEDOM AND LUXURY

The great Enlightenment author Voltaire (1694–1778) is often cited as a secular patron saint of freedom of speech, human rights, and tolerance. In spite of his spirited and courageous defense of these principles, however, Voltaire, like John Locke before him, was well aware of the limits within which he deemed them applicable— limits that were defined most of all in terms of money. Voltaire was a wealthy man who lived in castles, lent very substantial sums to aristocrats, and invested in overseas trade. He is a good example of the conflict between universal values and particular interests, and the way this conflict was resolved in the minds of those who found themselves on the winning side.

Voltaire's relationship with his rich clients mirrors a central dilemma of many Enlightenment thinkers. They needed connections in aristocratic circles for protection against arbitrary arrest or other professional dangers, and also in order to generate enough income to survive, but they also found themselves compromised by these connections. Even the atheist firebrand Denis Diderot was obliged to make the long journey to St. Petersburg in order to thank his patron Tsarina Catherine the Great. Traveling there in the honest hope of being able to move the monarch toward Enlightened reforms, he was soon disillusioned by the pragmatic Catherine, who was happy enough to discuss philosophy in an armchair by the fire, as it were, but who regarded practical politics as an entirely separate sphere.

Voltaire's most important relationship was with King Frederick II of Prussia, who praised him in effusive letters and also invited him to his court. Unlike Diderot in Russia, Voltaire was not disappointed by what he found in Potsdam, because he had not expected so much. He was amused by the provincial courtiers whom he loved to provoke, but he had no intention of even trying to convert Prussia into a kingdom of Enlightened equality. Voltaire was free from such sentimental illusions. In his famous novel *Candide*, he had warned that every sack of sugar from the Caribbean was stained with human blood, because it had been produced by slave labor, but he was nonetheless happy to invest his own money in the Compagnie des Indes, which realized immense profits by operating slave plantations for sugar, tobacco, and other goods. He even had a suave and cynical argument at the ready to defend his incongruous position: "We only buy negroes as house slaves. Some people reproach us for this trade. A people selling its own children [however] is even more worthy of condemnation than the buyer. This trade also shows our superiority; he who accepts a master was born to have one."

They only have themselves to blame, the philosopher decided while leafing through his accounts. Slavery was unfortunate, of course. He himself had written about it very movingly. But it was a real tragedy only if it happened to white people, because, he observed, "I see people, who seem far superior to the negro, just as negroes are superior to apes, and apes to oysters. . . ."

It was quite typical for Europeans in Voltaire's day to believe in the inequality of humans according to ethnic criteria, and to deduce from this that white people, meaning themselves, simply had more rights by nature—though the assertion was usually based on a near-complete ignorance of other cultures and other peoples. Neither Voltaire nor later thinkers such as Immanuel Kant or Friedrich Wilhelm Hegel had ever set foot on another continent, learned a non-European language, or studied other cultures. They still wrote with sweeping conviction about the inferiority of non-European peoples. Time and again, they reached the conclusion that white Europeans (and more specifically men, preferably highly educated ones like themselves) were somehow superior to the rest of humanity.

There were those among Voltaire's contemporaries, however, who were better able to identify those philosophical convictions that conveniently served their own personal advantage. It is a common mistake to condemn the past according to the morality of the present, but there were some people in the seventeenth and eighteenth centuries who did not believe in the superiority of European civilization. Perhaps even more surprising, there were also some who believed that the Europeans' superiority in engineering, military technology, and navigation did not entitle them to plunder and subjugate other peoples.

Voltaire may have been a child of his time, but while he was defending the rights and the ethics of slaveholders and colonists, his French compatriot Denis Diderot was writing, even if anonymously, an impassioned indictment of slavery and imperialism in his *Histoire des Deux Indes* (1770). The moral horizons of a period may be limited by custom and culture, but these horizons are movable, and it is the task of philosophers to move them.

The wit and often the nobility of tone with which Voltaire celebrates human rights can be admirably effective. All men are brothers, he wrote, children of the same God. Not all these children were equal, though. Nature might have made them in the same way, but society had other ideas: "It is impossible on our poor globe not to have all people living in societies divided into two classes: one of the rich, which commands, and the other of the poor, which serves."

Installed in an elegant chateau, Voltaire had no difficulty accepting this order, while he insisted on equality and freedom in his ethical writings. These, however, were little more than calendar wisdom, to be turned and twisted as the occasion demanded:

> At the bottom of his heart, every person is entitled to think himself the equal of all others. Does this mean that the cardinal's cook can command his master to serve him, too? The cook can say: "I am a person, just as my master. When I was born I cried, just as he did. He will die, like me, with the same torments, the same ceremonies. We both have to bow to our physical, animal needs. When the Turks conquer Rome, I will be cardinal and he will be my cook: he will cook my lunch!"

At this point, however, Voltaire reigns in his egalitarian ideas: "This speech is just and rational. But until the Great Turk comes and conquers Rome, the cook will have to cook, or human society is perverted."[1]

Voltaire was no revolutionary. He insisted that social problems and injustice were due to tyrannical excess and to superstition, to excessive power and excessive ignorance, to fanaticism. There was nothing wrong in itself, he concluded, with the rule of aristocracy and Church. As an educated man, he felt no obligation to believe the fairy tales told by the priests, but he was convinced that it was necessary for the broad majority of people to believe them. "Christianity is certainly the most ridiculous, absurd, and bloody religion, which has ever infected the world," he wrote in a letter to Frederick II, adding: "I don't tell that to the common plebs, which is not worth being enlightened and which will bear any yoke; I say this among men of honour, among men who think about things."

The common people, Voltaire believed, were morally and intellectually too feeble to live without religion: "Man always needs a connectedness, and even if it is ridiculous to sacrifice to fauns, forest gods and Najades, it is still more rational to pray to these fantastical images than to sink into atheism." *Écrasez l'infame?* Later, perhaps.

Unresolved contradictions and convenient compromises such as this did not make Voltaire a less interesting thinker in the eyes of his contemporaries. On the contrary—the party of Enlightened and reformist bourgeoisie made him a star, for despite their theoretical interest in human rights and social justice, they were anxious to crush all attempts at any real social change that would undermine their own position. Voltaire's eye for publicity, and for useful scandals, propelled him into the stratosphere of fame. He did, however, frequently divide opinions within the elite, being prepared to go far enough to be embraced by those who were more progressive, but not so far as to alienate his aristocratic and highly lucrative clients. Atheism seemed dangerously corrosive to Voltaire, and he fought it wherever he could with considerable elegance: "What is tolerance? It belongs to humanity. We are all full of weaknesses and errors: let's forgive one another our stupidities; that is the first law of nature."

Voltaire preferred to use his influence from his elegant abode in tolerant Switzerland. From there, he could safely take part in French intellectual life: in discussions, in ruining reputations, or in building up young talents—either to secure them as potential allies or to be able to knock them down more effectively later. He was a racist before "racism" was invented, an apologist of slavery who benefited financially from slave labor, a friend and banker of princes who ruled according to the very methods he so decried in his books. He was a great stylist, but also a self-serving cynic who argued that the mass of the people should not be educated. On the contrary, they should be kept docile through belief in what he regarded as fairy tales and the fear of divine punishment that they engendered.

As a young man, Voltaire had been forced to spend three years in exile in England—not because he had written a daring philosophical tract but because he had unwisely and very publicly questioned the sexual potency of an influential aristocrat, who had promptly had him thrashed by his servants and had threatened worse. This enforced sojourn across the channel had been a revelation to the young Voltaire.

Even if the cooking in England failed to impress him, he was deeply taken with the freedom of opinion practiced there, with the country's constitutional monarchy, the stock exchange (which he described as the country's true temple), and the pragmatism of the English. When he returned to the Continent, he was convinced that only economic growth could make societies more open, tolerant, liberal, and peaceful. Like Bernard Mandeville and later Adam Smith, he believed that individual greed could serve the common good. Merchants, not aristocrats, he felt, were the true heroes of society.

It is not hard to imagine how opinions such as these were received in absolutist France. Readers during our own time may notice parallels with twenty-first-century neoliberal economic thinking. Voltaire—the first neoliberal?

The resemblance is not coincidental. The ideology of the free market is an echo of the rationalist, deist Enlightenment propagated by Voltaire and his admirers. Both share fundamental premises such as the essential rationality of human beings, individual freedom, tolerance in the marketplace, the imperative of freedom from outside intervention, the self-regulating force at work within the economy, and the necessity of a meritocratic elite steering the political and economic fates of whole continents.

The legendary quips and putdowns from Voltaire's pen remain immensely quotable, as do his stirring words about human rights, dignity, and the necessity of reasoned argument. At the same time, his thinking is already a construct of convenient compromises. His tolerance is the tolerance of the marketplace, his ideas are destined only for a small elite, his concept of freedom holds the weak responsible for their lot, and the rationality he defends is to rule the masses with the weapon of superstition.

According to Voltaire, it is a waste of time to pursue an ideal world, the El Dorado he described in his novel *Candide*—a land in which everyone is happy without priests and dogmas, and with food enough for all. He himself, he felt, would have been bored there; only distinction made life interesting. The individual therefore has no choice, he believed, but to seek contentment in retreat from an imperfect world, as did Candide and his creator, who famously opined: *"Il faut cultiver notre jardin."* We find happiness not as fighters for freedom but as cultivators of flowerbeds—provided we have a garden.

INHERITED COMPROMISES

At the beginning of this book, I asked a straighforward question: What changes in a society when the climate changes? For the early modern period, it appears that the crisis of agriculture following environmental cooling accelerated a social and economic dynamism carried by a rising middle class, by stronger trade, empirical knowledge, expanding literacy, growing markets, and intellectual renewal. The result was a move from feudal to capitalist societies, from the fortress to the market.

If this great transformation so deeply affected societies four hundred years ago, what pressures for change are there now on the societies that have grown out of them? And, if today's wealthy Western societies are heirs to the Enlightenment, and to the contradictions between its claims and its practices, can and should these foundational ideals play a defining role in another great transformation? Or should they simply be discarded?

What pressures arise for such changes? Then as now, there is pressure from climate change on economic and social structures, on natural resources and social cohesion—forcing countless people to leave their homes, their families, their countries and thereby disrupting social structures and practices in place for generations. Then as now, a shift in weather patterns causes natural disasters, upending societies and creating fear, as well as exacerbating the need for change. Then as now, the world's most successful economies operate on a system of economic growth based on increasing exploitation of human and natural resources.

Here, however, is where the similarities end. During the seventeenth century, the cold winters and rainy summers were not commonly understood as part of a global climate event. As we have seen, they were attributed to divine anger, or to inexplicable local factors. There was no global climate model because the scientific method itself was only nascent. The responses to these developments therefore were also not systematic; they occurred as random answers to a problem that was not well understood. By the time climatic conditions reached pre-1570 levels again, sometime during the eighteenth or early nineteenth century, Europe's societies had been transformed.

This story seems to invite an optimistic parallel with the present, but such a parallel does not hold. It is true that early modern Europeans—or at least their elites—apparently profited from the Little Ice Age, and that this global climatic variation apparently rectified itself. Neither of these factors, however, holds true today.

Today's climate change comes largely from the emission into the atmosphere—within just a couple of centuries—of much of the

CO_2 stored in the earth over hundreds of millions of years. This process is not due to a natural variation in the activity of the sun or in solar spots or orbital variation of the Earth; it will not rectify itself. On the contrary, it is accelerating. There will be no moment when its effects have naturally disappeared to reveal societies that have grown not only in economic success but also in wisdom. *Homo sapiens* is a clever and fascinating primate, acting on impulse and desire, but we are no cleverer, and arguably have no more ultimate insight into our condition, than our ancestors did.

Unlike the early scientific thinkers of the seventeenth century, today's climate scientists can make evidence-based and detailed projections of climate developments. And while science by definition works with models, with different degrees of accuracy, these models are being borne out by a vast array of data collected as the changes manifest themselves around the planet—from single-cell organisms to the stratosphere.

Taken together, climate data and the projections based on them produce a series of scenarios ranging from severe economic and cultural challenges to biological and civilizational collapse. So far, predictions about the effects of climate change have come true— and much more rapidly than scientists had thought probable. The political, social, and cultural consequences of this phenomenon are as yet unknown, but they are likely to be sweeping.

Perhaps this is all too vast for us to take in, perhaps it is simply too inconvenient, perhaps it seems too abstract—but it becomes more concrete with every hurricane and every flood. The rich Western societies of today are no more effective in combating climate change than those that existed around the year 1600, albeit for different reasons. Historically, the idea of global climate change simply did not exist, and there were no data to support such a theory. Today, data and projections exist, but they point in a direction that is too frightening, and also simply too big, on an individual level, as well as too disruptive to national and global economies built on growth and expanding exploitation.

If you want to stay competitive, the reasoning goes, you simply cannot afford to take all this seriously. In the political battle between short-term economic interests and long-term climate demands, the former almost always wins, trailed by face-saving initiatives laudable in themselves, perhaps, but not powerful enough to compromise profit margins. Voltaire would have understood this perfectly—both as a playwright propagating humanist ideals and as a businessman with an eye firmly focused on the bottom line.

The Little Ice Age first hurt European societies economically; eventually it catalyzed change in other areas. Today, the economic costs of global warming are mounting: in environmental destruction, in loss of life and income, in disaster relief costing billions, in drastic losses of biodiversity and natural habitats, in pressure on marginal agricultural areas, in shrinking coastlines and vanishing Pacific islands, in droughts and hurricanes, and also in an enormous increase in migration from the most-affected areas. All these changes drive up the costs of continued growth and of growing exploitation as an economic model, to which, claim those who profit most by it, there is no alternative.

The pressure on the economic model of richer societies is already evident, and it could serve as the catalyst for the birth of a new kind of society whose shape and structure we cannot yet fully formulate. In contrast to 1600, this pressure is not occasional, and it is not likely to subside.

In telling and analyzing the story of the Little Ice Age, I have chosen Europe as an example of how such changes occur; an informed perspective today, however, can only be global. Herein lies another difference between the early modern past and the present: It makes sense to describe European societies during the seventeenth century without also looking at cultural or political developments in Asia or Africa. Economic, political, and cultural interaction was relatively limited and largely dominated by European powers.

Today, the consequences of Western lifestyles touch the remotest corners of the globe, while the consequences of social and natural disasters will be felt, tsunami-like, across oceans and continents. In an age that sociologist Zygmunt Bauman dubbed "liquid modernity," global interests, potentials, and problems can no longer be contained; every current produces a countercurrent. Today, it would be impossible to write about Western societies without understanding their complex though frequently still lopsided relationships on all continents.

And, as in the seventeenth century, we have our own contemporary debates about the enclosure of the commons, expropriations, and rural depopulation, as sociologist Saskia Sassen relates:

> One important explanation for these migration movements is extreme violence. Young people from Central America tell that they fled from violence in the cities, from gangs, but also from police violence. Another cause lies in 30 years of international development policy, which has left a great deal of dead land behind. Plantations, land grabbing, mines—all that has driven millions of people away from their own surroundings. But they are hardly acknowledged, because they are not officially recognized as refugees. Entire areas have been rendered uninhabitable. Climate change results in rising sea levels and desertification and further reduces the amount of usable land. Those concerned only have the option of moving into the great slums of cities, or, if they can afford it, migration.[2]

Viewed from a historical perspective, our present situation shows all the hallmarks of a prerevolutionary period: increasing asymmetries between the business models of many societies and the resources they depend on, between rich and poor, between movable capital and local populations, between conspicuous consumption and social solidarity. In societies actively undermining the foundations of their continued existence—both economically and philosophically speaking—the occurrence of some kind of dramatic collapse seems to be only a question of time. Some observers already claim to see the first signs of an irreversible cultural and democratic unraveling, even as ecological disasters are unfolding before our eyes.

What, if anything, can the seventeenth century tell us about the twenty-first? Perhaps it is not a question of parallels but of continuities. Understanding what the present has inherited and how it is dealing with its frequently ambivalent legacy can also suggest different ways of thinking, and of acting.

NEW METAPHORS

During the Little Ice Age, Europe found new metaphors for thinking about itself. At the end of the sixteenth century, climate change was still understood as divine punishment, part of the central relationship between God and his creatures. Based on this metaphorical understanding, the initial reactions were rational but futile: Processions, penance, sermons, rituals, and exorcisms did little to change the weather.

Only when this metaphor changed, when nature began to be perceived as a kind of clockwork, did empirical changes in agriculture, trade, and engineering begin to alter the impact of this cold spell on humanity. As part of this development, new ideas made it possible to conceive of new societies in which intellectual freedom would be more important than conformity, and the rights of the individual would be more important than the rights of the group. Those who saw not only the moral but also the social worth of these ideas fought the resistance to them, often with physical bravery, and then propagated them.

As nature began to be seen not primarily as the gift of a divine Creator but as a vast, interlinked mechanism, this change in metaphors also altered the shared transcendence that formed a society out of the people living in a single space. For several centuries, this transcendence, in Europe, had been overwhelmingly Christian. During the Enlightenment, it received a series of new, more abstract shapes: the Republic of Letters, the Empire of Reason, and later Progress, the Fatherland, the nation, the workers' paradise, the triumph of the master race, the people, the Republic.

As Pankaj Mishra emphasizes in his *Age of Anger: A History of the Present*, this struggle between metaphors, between ways of making sense of the world, resulted in real and often endemic violence. Every ideology comes with a set of values, an instruction manual of how to be and to behave in the world. The spread of colonialism and of capitalism was marked by great violence, as was the resulting backlash of revolutions and independence movements. Locked in a logic not so much of opposition but of rivalry, the dreams of one part of humanity invariably turned into another's nightmares.

Twenty-first-century climate change makes it a matter of urgency to rethink once more our cultural metaphors, as well as humanity's place within the greater scheme of things. Our success and failure in dealing with climate change will depend to a large extent on how societies shaped decisively by the Enlightenment deal with this heritage, and how the Enlightenment itself is understood.

What is today called "the Enlightenment" was never a philosophical school with its own catechism, but rather a landscape of broad and frequently antagonistic debates between believers and unbelievers, libertarians and statists, cynics and social reformers, materialists and anarchists, utopians and pragmatists. The historians and philosophers who became its official chroniclers were masters at clipping, mowing, pruning, weeding, hacking, polishing, and cutting the vast and wild panorama of Enlightenment debates until it resembled a manicured French park.

This acceptable, canonical Enlightenment privileges the rationalist thinkers, from Descartes to Kant and the Marquis de Condorcet. In this, we find stifled echoes of seventeenth-century debates, as different Enlightenment traditions were pruned down to something resembling a coherent and harmonious body of thought. Descartes and other rationalists argued that human beings are quintessentially rational and expansive and capable of infinite self-improvement. The task of the Enlightenment is, therefore, as Kant famously put it, "man's exit from his self-imposed immaturity." The task of the individual becomes to develop and expand the dominion of rationality and to work against the influence of irrational desires, instincts, and mere sentiment. Only as a fully rational being, making free and rational choices, can a person achieve his or her potential.

This reduced, purely rationalist version of the Enlightenment still informs present-day debates about liberty, equality, and solidarity, shaping a great deal of social, political, and especially economic thinking. It is therefore important to understand its underlying structure.

THE THEOLOGY OF THE MARKET

> The ideas of economists and political philosophers, both when they are right and when they are wrong, are more powerful than is commonly understood. . . . Practical men, who believe themselves to be quite exempt from any intellectual influence, are usually the slaves of some defunct economist. Madmen in authority, who hear voices in the air, are distilling their frenzy from some academic scribbler of a few years back.
>
> JOHN MAYNARD KEYNES, *The General Theory of Employment, Interest, and Money*

First and foremost, we need to acknowledge family resemblances. The rationalist Enlightenment owes its position not to its inherent truth but to its social energy and to the fact that it could be grafted onto an older, well-established intellectual tradition: Christian theology. There is a striking similarity between this Enlightenment concept of reason and the concept of the Christian soul. In each case, the immaterial and noble part of an individual demands that "lower" instincts, desires, and emotions be suppressed, so that a higher realm of purity and happiness can be reached.

It is not only reason that has powerful theological echoes. Consider the great triad of French Enlightenment values: *liberté, égalité, fraternité.* The central idea of freedom obviously builds on the Christian dogma of free will, which is central because it is necessary for redemption. Without free will, there can be no possibility of sinning; without sin, there need be no forgiveness and no redemption, and hence no Christianity.

The assertion of human equality reflects a powerful strand of Christian thought: exhorting believers to love their neighbors and see the person of Christ in the lowest and weakest people. Historically, though, this was very quickly tempered by interpretations of who was to be regarded as an equal and who was too low, too alien, too heathen, or too economically useful to be regarded as a potential equal. *Fraternité*, or *solidarity* in modern terminology, has often been carefully redefined before being applied.

Another central aspect that is not part of the French revolutionary creed is the Enlightened belief in progress, another echo of Christianity. Faith in historical progress mirrors the carrying-out of God's will, while the Enlightened love of utopias and ideal societies builds on Christian hopes of a better future after the Last Judgment, when a just and divine order will finally rule.

It was relatively easy to introduce this kind of Enlightenment into societies already saturated with religious images and rhetoric. All that was necessary was a change of vocabulary. Where once God and Scripture had been the ultimate arbiters, Nature and Reason now stepped into that role. The structures of the argument remained the same, grafted onto an older tree and nourished by its energy.

With the structures of beliefs remaining largely intact, it was easy for the justifications for economic and political power to remain so, too. Christian rulers had conquered and oppressed others in the name of their true faith, and Enlightened businessmen and politicians felt entitled to do the same in the name of rationality and progress. Colonial subjects, the working poor, criminals, women— all were deemed less rational, and many a scientific career was made by measuring skulls and weighing brains in order to prove that these differences had an objective biological basis.

After 1945, and again after 1989, these manifestly theological traditions of thinking about human beings and their societies appeared to have foundered disastrously. The great messianic dreams of Fascism and Communism had proved to be murderous illusions. Only liberal democracy remained to offer some kind of transcendence, but with its constant inherent compromises, with each voice carrying the same weight as the next, it seemed to some that it had merely enshrined a banal mediocrity.

A new kind of messianic hope was now introduced, a different kind of religion, a new theology. It was less created than—appropriately for such a religious idea—resurrected from the dead, from the seventeenth and eighteenth centuries. It was the gospel of the unimpeded rule of the free market.

Relying on the free market rapidly became the only game in town for governments, corporations, and international institutions, including almost every university economics department in the world. Individuals who did not or would not accept its absolute power were sidelined in academia and other areas of influence. The idea that humanity had finally found a fact-based, statistically proven, and empirical principle on which to build and run a global society was simply too seductive.

Like the Enlightenment, however, this economic theory was also an expression of the social and economic interests of a particular group. Like its predecessor, it also drew its force, its familiarity, and its plausibility from being grafted onto the same old tree of Christian theological thinking, as mediated by the canonical thinkers of the Enlightenment.

What Joseph Stiglitz calls "market fundamentalism" bears some of the same religious hallmarks as the officially sanctioned rationalist Enlightenment. Both emphasize the importance of reason, freedom, equality, the capacity of infinite self-improvement, individualism, and a messianic orientation toward a better future in which the contradictions and asymmetries of politics are resolved through competition, supply and demand, and rational self-interest.

All of these assertions are counterfactual, or at the very least debatable. Humans are largely driven by the same instincts as other animals; their rationality, that echo of the immortal soul, sits atop, and cannot completely control, the animal passions below. Humans often do not act out of rational self-interest at all. Their fundamental motivations are, as so many seventeenth-century thinkers knew, driven not by profit maximization but rather by sex, fear, and a striving for recognition. In addition, free will remains a problematic idea in the face of genetics, epigenetics, behavioral and evolutionary psychology, and even sophisticated modern advertising. Equality may exist as an idea, but there is hardly a place on earth where it is not eroded by powerful patterns of exploitation and discrimination. In an economic context, it cannot be said to exist at all.

It is instructive to see how reliant economic theories are on philosophical claims, and how profoundly bad philosophy can compromise those theories from the outset. After rationality, freedom, and equality collapse, so too do other axiomatic assumptions—the ascent to a perfect market in which freedom reigns supreme, and the reliance on an invisible mechanism that magically produces justice and plenty.

This now-outdated neoliberal concept of how a human economy works does not begin with an understanding of human nature or social structures or goals. It elides complex motivations and constraints into a posited rational self-interest. Flying in the face of all evidence, it assumes that transactions in the marketplace happen on a free and equal footing, that both sides have the same amount of choice and information. No social reality throughout history supports that claim. It is inherently counterfactual, constructing what is in effect a theology of the market.

As a body of teaching that traces its ancestry via the rationalist Enlightenment back to biblical thinking, neoliberalism also reproduces a biblical relation to nature: Subjugate the earth. Humanity's relationship with nature becomes determined by exploitation; nature must be subservient to human goals. Its integrity is of no importance, its beauty impossible to quantify and therefore nonexistent; it is a source of economic wealth, not wealth in itself.

This theology of the market was a response to the dramatic breakdown of the political, militant ideologies of the early twentieth century. Like any other successful principle of social and symbolic order, however, it also had to offer some kind of transcendence to the inhabitants of market societies.

Transcendence used to be in heaven, but the Enlightenment converted it into an abstraction—*liberté, égalité, fraternité*. It enters the free market first and foremost through the act of consumption—the moment of personal transformation, of membership in a commercial tribe, of becoming one with the ideal, of touching the divine. Just as the Church represented saints in extravagant and sublimely effective works of art, the market generates advertising to communicate its own transcendent realm of wealth, coolness, youth, authenticity, masculinity, femininity, sexiness, and self-affirmation, all of which can be acquired, supposedly, via a simple money transaction.

Traditionally, personal and social identity was overwhelmingly circumscribed by the circumstances of one's birth. Everyone was born into a particular religion, region, class, trade, tradition. Most people never escaped this rigid structure. Since the 1950s, or thereabouts, we have been accustomed to regarding that structure as loose at the joints, porous, even breakable. You move away from home, you meet other people, see other things, you marry out of your caste and even race, and you or your children become a different kind of people entirely. Within a single generation, the fate and status of individuals and their families can alter dramatically. Even acts of consumption and brand identification can become defining moments for the individual.

This is a giant step from the social reality of the seventeenth century, though the vastly increasing income inequality of recent decades has entailed the return of some of those ancient habits: dependence, immobility, lack of education, inability to rise socially. The mercantilist idea of economic growth based on exploitation is still dominant in Western economies, though it may go by another name. While it fueled the global dominance of Europe, and later North America, for more than four centuries, it has now become an existential threat, as the price of the increasing consumption of natural resources such as fossil fuels becomes clearer. And yet, there is still a solid stratum of opinion that insists the market will solve all problems, that nature is a free resource to be exploited, that humans are essentially rational, free, and equal in the marketplace, and that the arc of free-market capitalism bends toward peace and justice.

THE MARKET AND THE FORTRESS

The great transformation of European societies during the Little Ice Age can be described as a move from the Fortress to the Market— from communities centered on castles, keeps, and churches to more diverse ones in which exchange and trade, cities and literacy were more crucial than belonging and faith.

After the fall of the Iron Curtain, it appeared that the market had achieved a historic and irreversible victory. But if what we have now is victory, then it looks very different from what its prophets had assumed. The introduction of free markets, it transpired, does not entail an automatic strengthening of democracy, liberty, and equality for all. On the contrary.

In 1944, Karl Polanyi analyzed the way in which markets shape societies and warned that one day societies could be degraded to being "a mere adjunct of the market." His astute observation has already become reality, as the tentacles of market interest penetrate lives more deeply and more relentlessly than ever before, and personal identity now becomes a matter of aggregate consumer decisions.

The market as a model for society can stand for openness, toler-ance, exchange, equality before the law, constructive competition, cosmopolitanism; it can be synonymous with the Agora, the market square in ancient Greece where equal freemen did business, debated, and lived. It becomes different, however, if profitability becomes the only key to understanding the world, the only valid criterion. Then it becomes a cold place for most people forced to live without pro-tection, to adapt, and to play a game stacked against them—a game designed to transfer wealth from the many to the few.

This Janus-headed nature of market societies between Agora and *Hunger Games* has vast implications for democracies. Democracy needs the openness and exchange of the marketplace to thrive, but robust market success does not require democracy—witness China and Singapore. In this very lopsided relationship, democracy is in danger of being chewed up and spat out by business interests, or transformed, as in Russia and even the United States today, into a submissive rubber-stamping mechanism, enacting laws that cater to powerful corporate interests.

During the seventeenth century, this was exactly how the mercantilist state was supposed to function. The concept of the common good was debated by philosophers in their libraries or their places of exile, but it did not influence political priorities. The purpose of the state was not to deliver the best possible quality of life to the greatest number of people, but instead to maintain and enlarge the power of the monarch through economic growth, export, exploitation, and war.

Mercantilist markets, however, were different from today's markets in that they were understood to be no more free than any other aspect of societal life. To the rulers and thinkers of the sixteenth and seventeenth centuries, the market was not freer than the military and the wars it fought. It was commanded, steered, overseen, and facilitated by the monarch and those acting on his behalf.

The idea of a "free market" is a tempting but mistaken conflation of economic ideas and Enlightenment rhetoric. Of course, no market could ever be free in any meaningful sense of the word. To be able to function at all, markets depend on regulation and oversight, on courts of law and legal frameworks to make contracts enforceable, on infrastructure and social priorities, on roads and ports and schools. Markets are an expression of social activity, not a self-contained realm existing apart from it. Belief in a "free market"—one somehow capable of not only regulating itself but also steering humanity toward a moral goal—is as deeply theological as the belief in divine providence.

Bernard Mandeville was one of the first to equate the idea of a free market in which the individual followed his or her own venal desires, and the network of these desires achieved a common good.

He was ahead of his time. As an economic model, this dismal vision of society inspired the *laissez-faire* capitalism of early modernity. The resulting social reality in much of Europe supplied the ideas and the energy behind a long century of revolutions, from 1789 to 1917, and from there onward in anticolonial revolts, which finally demanded the promises of the Enlightenment for all. The ideologies inspiring or emerging from these revolutions—from Communism to Fascism and Rhenish capitalism—all possessed the same transformative energy. They also shared, in principle at least, a strong impetus toward the improvement of material conditions for the majority.

It is striking how strongly the bleak social and economic ideas of the seventeenth-century mercantilists have made a historic comeback since the collapse of the Soviet Union. Even in the richest societies, rising poverty and increased inequality of income and opportunity indicate that acceptance of "beggar thy neighbor" policies has made a successful return.

Seventeenth-century economies and technologies did not have the breadth of penetration to inflict lasting and profound damage on the ecosystem as a whole. But by the 1950s, the technological reach of humanity had increased so much that it was beginning to interfere with the natural environment. Scientists are still debating the idea that humanity has ushered in its own geological age—one they have dubbed the *Anthropocene*. But while arguments are still being exchanged, there can be no doubt that various technologies, and the numbers of people making use of them, are transforming our natural environment.

The intensifying exploitation of natural resources has triggered a series of possibly irrevocable changes whose complex consequences are not yet fully understood. Included among them are the collapse of entire ecosystems, the extinction of species, and the poisoning and degradation of vast areas of the globe. Proxy wars, toxic waste disposal, and the cynical installation or support of cooperative dictators have added to the effects of climate change to create vast numbers of refugees and other migrants.

This cascade of exploitation is part and parcel of continued economic growth—without which, we are told, Western societies and also developing economies will fail and slide into disaster. Most of this growth, it is now clear, has in recent years been flowing into the same handful of coffers, though this has not significantly lessened the calls for more of it. Thus far, partly because of low investment, renewable energies cannot provide viable substitutes for fossil fuels, at least at current levels of economic activity and consumption in industrialized societies. The situation is stark: We are trying to stay ahead of economic failure by destroying the natural environment we need to flourish in—perhaps even just to live and breathe in. The doctrine of economic growth based on exploitation—the legacy of the seventeenth century—has become a mantra of collective suicide.

In Europe and its satellites, the rise of a literate, educated, and entrepreneurial middle class created a world of markets, of democracies, and of ideas such as individual freedom and basic, inalienable human rights. The legacy of this social transformation, however, has been the struggle between interests and ideas, profits and principles. If it was always indifferent to any concept of moral progress, it has only recently, and unapologetically, revealed itself to be so. In the competition between ideas of a common good and the drive for private wealth, the latter has surged ahead. Our contemporary societies are edging closer to the world of Thomas Hobbes than to that of Spinoza.

Today there are millions in Europe itself and in other parts of the rich world who regard themselves as victims of market forces. Some of these are beating a retreat into a new kind of Fortress—national, religious, racial—with its own clear and impregnable borders.

The Fortress offers security, control, a sense of self, a sense of dignity—a stable identity. In any Fortress, the inhabitants are all cut from the same cloth. As far as they know, they share the same historical experience, the same culture, the same religion, and, if possible, the same ethnicity. They remain convinced that liberty cannot be valid for everyone, that equality is an illusion, and that solidarity should be directed toward one's own kind.

Just as the thinking of the Market has its founding fathers, the Fortress has an intellectual canon of its own. Following Jean-Jacques Rousseau, those who choose the Fortress believe that the Enlightenment was a historic mistake—or a failure. Following Hegel, they are convinced that their own group has a manifest and glorious historical destiny, that others are too inferior or just too different to be afforded the same rights that they themselves possess. A strong, high wall appears to be the answer to their problems. Their Fortress is a place of refuge as well as resistance, sometimes armed, to outside forces.

The Fortress and the Market once represented stages of historical development, but today they serve as ideological alternatives.

How present climate change will affect today's societies will largely depend on whether the responses to it are essentially theological or evidence-led. The gospel of infinite growth, productivity, and innovation teaches that all problems will eventually be overcome through market mechanisms. But what if they won't? What if the rate of growth that advanced economies require to maintain growth produces a fatal degree of environmental degradation? Can our societies remain stable without economic growth? When do we say it is time to follow the scientific evidence? Is there an imaginable point at which societies decide that their priority must be to rely not on invisible friends or invisible hands but on the best available scientific thinking based on evidence that is there for all to see—and breathe.

There must be hope. The Enlightenment encompassed a range of opinions and arguments, but its official face was quickly narrowed to represent ideas compatible with theological ways of thinking that were already prevalent in society and were useful to those in power. This kind of thinking—from Christian theology to the rationalist Enlightenment and finally to the fiction of the free market—has brought great material prosperity to many, if not to all, but it is now approaching its limits.

What of the nontheological Enlightenment, the ideas put forward by those who had to flee or were executed, and who were eventually marginalized or excised from the history books? What about the Enlightenment of Lucilio Vanini, Giordano Bruno, Baruch de Spinoza, or Pierre Gassendi? What difference would it make if human beings were seriously to consider themselves part of a material universe, of an evolved natural world, of the ecosystem? What if human societies banished any thoughts of subjugating the Earth and of intervention from beyond and concentrated entirely on the possibility of well-being or even just survival in a material universe? What if we gave up thinking of ourselves as the rulers of nature and instead regarded ourselves as a species of interesting primates dependent on their natural environment, just like any other species?

The Enlightenment has not failed. It has merely been hijacked and castrated. In contrast to the thinkers usually highlighted in histories of philosophy, its most original exponents tried to rethink the world and humanity's place in it as a space without theology or providence. Time and again, these voices were silenced, but this marginalized tradition of Enlightened thinking reaches far into the past, well beyond Lucretius and into antiquity. It conceived of human beings not as inherently rational and capable of infinite improvement but as passionate animals directed toward nonrational goals, yet endowed with greater imaginative and deductive faculties than their closest animal cousins. Correspondingly, these thinkers did not construct happiness as the self-actualization of rationality, or the greatest possible economic success, but as the fulfillment of desire.

What changes in society when the climate changes? This book is a work of history, not prophesy, but it is possible, perhaps likely, that the current economic and political principles of highly developed societies—growth and exploitation—will result in their decline or even collapse. This would occur precisely because their basic assumptions are theological in nature, and they do not think it necessary to take human control of the situation they have—once unwittingly, now all too obviously—brought upon themselves.

Subjugating the Earth was an ambition worthy of a small society of Bronze Age cattle herders and farmers, whose knowledge of the world extended scarcely farther than their eyes could see. Today, our understanding of the physical environment has changed radically. Our appreciation of humanity's place in this vast and vastly complicated system must also change, and we must accept that there is no invisible hand that will save humanity from itself.

It may be salutary to recall just how young and fragile are such ideas as human equality, democracy, and human rights. The Enlightenment itself struggled to life barely ten generations ago; until just two generations ago, there were still countries in Europe where women could not vote. Disenfranchisement on racist grounds was practiced even more recently. In a quasi-disguised form, it continues in some parts of the United States today. In terms of equal access to the ballot box, to education, to health care, to justice, and to the levers of power, democracy and human rights are nowhere near fully realized, even in some highly developed countries.

Liberal democracy is not, as many of Hegel's latter-day disciples would have us believe, a necessary consequence of historical progress. Instead, it is a largely accidental, contingent, and vulnerable historical experiment with an open outcome, revealed by recent developments to be in present danger of being subverted, ignored, left to atrophy, or eliminated completely. Democracy was born out of ideas first broadly debated during the Little Ice Age. It could easily die or be hollowed out to a mere façade during our own era of climate change, as living conditions for ordinary people become harsher and the very rich take more power for themselves.

Today, as the first palpable consequences of global warming result in greater migration and more natural disasters, a climate of fear is spreading throughout Western societies. Borders and minds are closed, walls and fences raised, but consumption intensifies while the exploitation of natural and human resources continues to expand. Distraction becomes a ubiquitous necessity.

Yet the liberal dream of the market risks being asphyxiated by its own overheating ambition. The wax in Mandeville's beehive is beginning to melt. The animals notice this, become worried, and redouble their efforts—flying faster, accumulating more, multiply-

ing, building new honeycombs as defenses against intruders, buzzing angrily and in increasing agitation. The whirring of millions
upon millions of tiny wings heats up the air. The bees know that
their world cannot last forever, but they still want more of what they
have now—they are bees, they cannot escape their nature. Soon the
beehive will be uninhabitable, food supplies will run low, neighbor will turn against neighbor: A whole swarm will be engaged in
a nihilistic fight for survival. And while the insects are generating
their own downfall, the beekeeper is nowhere to be seen. The bees
go on buzzing, beating their restless wings. They are bees. They
have no choice.

ACKNOWLEDGMENTS

VIENNA, CHRISTMAS EVE, 2017. The temperature outside is eight degrees Celsius. In spite of countless Christmas Markets serving mulled wine, it is almost like spring.

When I was finishing this book, I overlooked the sun-drenched hills of Los Angeles from a handsome office that the Getty Research Institute very kindly put at my disposal. Only a year afterward, the beautiful scene I enjoyed was scorched and charred by the devastating Southern California fires, which also forced the Getty Center to close down entirely for a few days.

Several months after delivering the manuscript to my German publisher, I began working on this translation. No doubt my concentration on the matter created what the French so charmingly call a *déformation professionelle*, but I could not help noticing the irony of translating a book on the Little Ice Age during one of the hottest summers Vienna has ever experienced, sitting at my desk in swim shorts and drinking ice water.

This book began as an attempt to understand what climate change is going to mean to our societies. Past performance may not be a guide to future performance, but it seemed interesting to me to try to write a case study detailing the importance of the impact of climate change in at least one global region, during one period. I might not have had the courage to undertake this without the encouragement of my agent Sebastian Ritscher, my wonderful editor Tobias Heyl, and my publisher Jo Lendle of Carl Hanser Verlag, who all supported this idea from the beginning.

Many private conversations and discussions with colleagues have given this book its present form. My thanks go out to many friends who patiently listened to my overly enthusiastic early attempts at shaping my material, and who helped me find this shape by asking the right questions, pointing me in new directions, or simply allowing me to see their responses. The most important for this project were, in alphabetical order: Thomas Angerer, Reinhold Baumstark, Toni Bodenstein, Philippe Buc, Thomas Gaehtgens, Jonathan Israel, Kenan Malik, Pankaj Mishra, Anthony Pagden, Philipp Roessler, Rainer Rosenberg, Alexa Sekyra, and Paul Verhaeghe. Finally, a special vote of thanks goes out to Markus Hoffmann, my agent in New York, and to my wonderful new American publisher, Robert Weil.

During my last months of working on this translation, I was privileged to enjoy another fellowship, this time at Vienna's Institute for Human Sciences, a most stimulating environment for scholarship as well as academic and public debates and discussion. I would especially like to thank Shalini Randeria, rector of the institute, for extending the invitation to me, as well as the staff and fellows who made my stay immensely enjoyable.

Veronica, my wife and soulmate, has very probably learned more about the Little Ice Age during these months than she ever thought she would—or wanted to. Still, she was always there, with encouragement, criticism, unexpected observations, unexplored references and resources, and endless cups of tea. Thank you, my darling.

NOTES

1. Hesiod, *Works and Days*, 126–201, quoted in: *The Oxford Book of Greek Verse in Translation*, ed. T. F. Higham and Cecil Bowra, trans. Jack Lindsay (Oxford: Clarendon Press, 1938), p. 135.

Prologue: Winter Landscape

1. John Dryden, *King Arthur* (London: Chappell, 1843), pp. 30–31.

"God Has Abandoned Us": Europe, 1570–1600.

1. I. H. van Eeghen, ed., *Dagboek van broeder Wouter jacobsz* (Gualtherus Jacobi Masius) prior van Stein: Amsterdam 1572–1578, and Montfoort 1578–1579, Groningen, 1959/1960, p. 4.
2. Ibid., p. 17.
3. Ibid., p. 75.
4. Ibid., p. 183.
5. Ibid., p. 280.
6. Daniel Schaller, 1595, quoted in M. Jakubowski-Thiessen, H. Lehmann, S. Schilling, and R. Staats, eds., *Jahrhundertwenden: Endzeit- und Zukunftsvorstellungen vom 15. bis zum 20. Jahrhundert* (Göttingen, 1999), p. 152, trans. PB.
7. Ibid., pp. 155, 157.
8. Thomas Rörer, *Predigt über Hunger- und Sterbejahre, von einem Diener am Wort*, 1571, fol. 47v.
9. See A. Fletcher and J. Stevenson, *Order and Disorder in Early Modern England*

(Cambridge: Cambridge University Press, 1987), p. 232; John Bohstedt, *The Politics of Provisions* (Farnham, UK: Ashgate Publishing, 2013), p. 27.

10. Francis Bacon, "Of the Vicissitude of Things," in *Essays* (New York: Little Brown, 1887), p. 295.

11. William Shakespeare, et al., *The Booke of Sir Thomas Moore*, ca. 1601–4, British Library, Harley MS 7368.

12. See Robert J. Knecht, *The French Religious Wars 1562–98* (Oxford: Osprey Publishing, 2002), p. 91.

13. *La Ciudad de Ariquipa*, p. 1061 (Copenhagen: Det Kongelige Bibliotek).

14. George Percy, *Observations by Master George Percy*, 1607, in American Journeys Collection, Document No. AJ-073, Wisconsin Historical Society, 2003, pp. 9–10.

15. Ibid., p. 20.

16. Ibid., p. 21.

17. *Heinrich Bullinger Diarium (annales vitae) der Jahre 1504–1574* (Basel, reprinted 1904), p. 104.

18. Manfred Groten, ed., *Hermann Weinsberg (1518–1597: Kölner Bürger und Ratsherr* (Köln: Studien zu Leben und Werk, 2005), pp. 293–300.

19. For a presentation of these data, see Valerie Daux, Inaki Garcia de Cortazar-Atauri, et al. "An Open-Access Database of Grape Harvest Dates for Climate Research: Data Description and Quality Assessment," in *Climate of the Past* 8 (2012): 577–88.

20. Niederösterreichisches Landesarchiv, Ständische Abteilung (NÖLA, S.A.), Landtagshandlungen ungebunden 1592–93, Landtag November 1593, Stellungnahme der Stadt Wien und der anderen mitleidenden Stadte und Märkte zur kaiserlichen Landtagsposition. Quoted after Landsteiner, "Wenig Brot und saurer Wein," in Behringer et al., *Cultural Consequences of the "Little Ice Age"* (Cambridge, UK, and Malden, MA, 2003), p. 137.

21. Landsteiner, p. 138.

22. Christopher Black, *Early Modern Italy: A Social History* (London: Routledge, 2002), p. 27.

23. Miguel de Cervantes, *The History of Don Quixote*, vol. 1, trans. John Ormsby (London, 1885), p. 392.

24. Quoted in Dietmar-Henning Voges, *Nördlingen seit der Reformation: Aus dem Leben einer Stadt* (Munich: C. H. Beck, 1998), p. 64. Trans. PB.

25. Ibid., p. 65.

26. Richard van Dülmen: "Die Dienerin des Bösen—zum Hexenbild in der frühen Neuzeit," in *Zeitschrift für historische Forschung*, 18. Band (1991), pp. 385-98, Trans. PB.

27. Quoted in K. von Zittwitz, *Chronik der Stadt Aschersleben: Mit einem Grundriß der Stadt* (Aschersleben: Carl Lorleberg, 1835), p. 159. Trans. PB.

28. Quoted in Thomas Kaufmann, *Konfession und Kultur* (Tübingen: Mohr Siebeck, 2006), p. 420. Trans. PB.

29. Christoph Schorer, *Memminger Chronik oder kurze Erzählung vieler denkwürdiger Sachen* (Bayrische Staatsbibliothek MS), p. 111ff.

30. Johann Georg Sigwart, *Drey Predigten von Dreyen Vnderschiedlichen Hauptplagen und Landstraffen* (Tübingen, 1611), p. 111ff.

31. Lucretius, *On the Nature of Things*, trans. William Ellery Leonard (New York: E. P. Dutton, 1916).
32. John Dee, *The Private Diary of Dr. John Dee* (London: Camden Society, 1842), p. 46, passim.
33. Michel de Montaigne, *Essays*, trans. Charles Cotton, ed. William C. Hazlitt (London, 1877), p. 333.
34. Ibid., p. 671.
35. Ibid., p. 437.
36. Ibid.
37. Ibid., p. 421.
38. Ibid., "On Cannibals," p. 137.

The Age of Iron

1. Johannes Clusius, letter, May 1601, BnF, Ms. 699, f 139. Marie-Elisabeth Boutroue, Koninklijke Nederlandse Akademie van Wetenschappen, Amsterdam, in Carolus Clusius, *Towards a Cultural History of a Renaissance Naturalist*, ed. Florike Egmond, Paul Hoftijzer, Robert P. W. Visser (Amsterdam: Koninklijke Nederlandse Akademie van Wetenschappen, 2007).
2. Ibid., Høyer to Clusius, April 9, 1593.
3. Thomas More, *Utopia*, ed. George M. Logan and Robert M. Adams (Cambridge: Cambridge University Press, 1989).
4. Quoted after Henry Kamen, *Early Modern European Society* (London: Psychology Press, 2000), p. 154.
5. Karl Polanyi, *The Great Transformation: The Political and Economic Origins of Our Time*. 2nd ed. (Boston: Beacon Press, 2001), p. 46.
6. Ibid., p. 45.
7. Ibid., p. 44.
8. Ibid., pp. 68–69.
9. Ibid., p. 67.
10. Denis Diderot and Jean d'Alembert, *Encyclopédie des arts et métiers*, vol. XIII (Paris, 1765), p. 4.
11. Quoted in Geoffrey Parker, *Global Crisis* (New Haven, CT: Yale University Press, 2013), p. 571.
12. Edward Chamberlayne, *The Second Part of the Present State of England: Together with Divers Reflections upon the Antient State Thereof* (London: John Martyn, 1676), p. 43.
13. Wouter Jacobszoon, *Dagboek*, quoted in A. Th. van Deursen, *Plain Lives in a Golden Age* (Cambridge: Cambridge University Press, 1991), p. 141.
14. Deursen, p. 142.
15. Jean Nicolas de Parival, *Abrégé de l'histoire de ce siècle de fer* (Brussels: François Vivien, 1660), p. 12.
16. Ibid.
17. Renward Cysat, "Von dem grossen und erschröklichen Erdgidem, so sich allhie ze Lucern, wie ouch in aller umbligender Landschat, und in anderen privinzen tütscher und welscher nation wytt und breit erzeigt den 18 Sep-

tembris des 1601 Jars," in *Der Geschichtsfreund: Mitteilungen des Historischen Vereins Zentralschweiz,* Band 3 (1846), p. 114ff.

18. Renward Cysat, "Collectanea," quoted in Christian Pfister, *Klimageschichte der Schweiz 1525–1860,* vol. II (Bern: Haupt, 1984), p. 94.

19. Quoted in Albert Fischer, *Das deutsche evangelische Kirchenlied des 17. Jahrhunderts,* ed. Wilhelm Tümpel, Gütersloh, 1905–1916, vol. III, no. 109, verse 1.

20. *Paul Gerhardts Sämtliche Lieder* (Zwickau: Verlag und Druck von Johannes Herrmann, 1906) p. 906.

21. Roger Boyle, quoted in Geoffrey Parker, *The Military Revolution* (Cambridge: Cambridge University Press, 1996), p. 16.

22. Marc Bloch, *Les caractères originaux de l'histoire rurale française* (Paris, 1960), p. 107; see also Guy Bois, *Crise du féodalisme* (Paris: Presses de Sciences Po, 1976).

23. Quoted in Parker, *Global Crisis,* p. 77.

24. Thomas Mun, *England's Treasure by Forraign Trade or, The Ballance of Our Forraign Trade Is the Rule of Our Treasure.* Written by Thomas Mun of Lond. Merchant, and now published for the Common good by his son John Mun of Bearsted in the County of Kent, Esquire (London: Printed by J. G. for Thomas Clark, 1664), p. 3.

25. Ibid., p. 8.

26. Ibid., p. 33.

27. Ibid., pp. 27–28.

28. Ibid., p. 34.

29. Jan Romein and A. Romein-Verschoor, *Erflaters van onze beschaving. Nederlandse gestalten uit zes eeuwen, vol. II* (Amsterdam: Querido, 1938–40), p. 281.

30. Ibid., p. 272.

31. René Descartes, *Oeuvres,* 1903, p. 466, Letter DLXXXIV, Soultrait, Bib. Bonna XVII, p. 94.

32. Quoted in Wilhelm David Fuhrmann, *Leben und Schicksale, Geist, Character und Meynungen des Lucilio Vanini* (Leipzig: Johann Gottfried Graffé, 1800), p. 277f.

33. Ibid., p. 307.

34. Ibid., p. 314.

35. Ibid., p. 316.

36. Ibid., p. 319.

37. Ibid., p. 322.

38. Quoted after Namer, *La vie et l'oeuvre de J. C. Vanini* (Paris: Vrin, 1980), p. 174.

39. Marin Mersenne, *L'Impiété des déistes* (Paris: Pierre Billaine, 1624), p. 512. Trans. PB.

40. Hugo Grotius, *Annales et histoires des troubles du Pays-Bas* (Amsterdam: L'Imprimerie de Jean Blaeu, 1662), p. 368. Trans. PB.

On Comets and Other Celestial Lights

1. Quoted in Howard Robinson, *The Great Comet of 1680: A Study in the History of Rationalism* (Press of the Northfield News, Harvard University, 1916), p. 24.

2. Johann Bödiker, *Christlicher Bericht von Cometen* (Cölln an der Spree [Berlin], 1681), p. 3.
3. Rev. Robert Law, *Memorialls; or, The Memorable Things That Fell Out Within This Island of Brittain from 1638 to 1684* (Edinburgh: Archibald Constable and Co., 1818), p. 170.
4. Christiaan Huygens, *Oeuvres Complètes de Christiaan Huygens*, vol. VII (Den Haag: Martinus Nijhoff, 1908), p. 312.
5. Pierre Bayle, *Various Thoughts on the Occasion of a Comet*, trans. Robert C. Bartlett (Albany: State University of New York Press, 2000), p. 13.
6. Ibid., p. 18.
7. Ibid., p. 19.
8. Ibid., p. 62.
9. Ibid., p. 30.
10. Ibid., p. 200.
11. Quoted in Steven Nadler, *Spinoza: A Life* (Cambridge: Cambridge University Press, 2001), p. 120.
12. Bayle, *Pensées diverses, écrites à un docteur de Sorbonne* (Rotterdam: Leers, 1683–1694), p. 45.
13. Quoted in Jonathan I. Israel, *Radical Enlightenment* (Oxford and New York: Oxford University Press, 2001), p. 161.
14. Spinoza's reply to his friend Oldenburg has been lost, but the messianic fever was continuing to spread. Jewish communities in Amsterdam, Italy, and the Habsburg provinces of Galicia and Moravia were seized by the enthusiasm, and even non-Jewish scholars began to accept Zevi's claim to divinity. Not all, however, were convinced. Two Talmudic scholars from Lemberg (now Lviv, Ukraine) began the long journey to visit Zevi and to see the Messiah for themselves. They were not won over by him, perhaps because they were adherents of a rival prophet. Still, the hope of imminent salvation had gripped entire groups such as the Jewish community of Avignon, who were making concrete preparations for their relocation to an earthly paradise.
15. John Evelyn, *The Diary of John Evelyn*, with introduction and notes by Austin Dobson (London: Macmillan, 1908), pp. 355–56.
16. John Locke, *Second Treatise of Government*, ed. C. B. Macpherson, p. 52.
17. Ibid., p. 9.
18. Bernard Mandeville, *The Virgin Unmask'd: Or, Female Dialogues Betwixt an Elderly Maiden Lady, and Her Niece* . . . (London, 1709), p. 3.

Epilogue: Supplement to *The Fable of the Bees*

1. Voltaire, *Dictonnaire philosophique, portatif* (London, 1765), p. 174.
2. Antje Lang-Lendorff, "Flüchtlingskrise. 'Wir sind auf diese Veränderungen nicht vorbereitet,'" Interview with Saskia Sassen, *Berliner Zeitung*, April 20, 2016. Trans. PB.

BIBLIOGRAPHY

Abel, Wilhelm. *Agrarkrisen und Agrarkonjunktur: e. Geschichte d. Land- u. Ernä-hrungswirtschaft Mitteleuropas seit d. hohen Mittelalter.* Hamburg and Berlin: Parey, 1978.

———. *Agricultural Fluctuations in Europe: From the Thirteenth to Twentieth Centuries.* Abingdon, UK: Routledge, 2013.

Acot, Pascal. *Histoire du climat.* Paris: Perrin, 2003.

Allen, Don Cameron. *Doubt's Boundless Sea: Skepticism and Faith in the Renaissance.* Baltimore, MD: Johns Hopkins Press, 1964.

Allen, Robert C. "Tracking the Agricultural Revolution in England." *Economic History Review* 52, no. 2 (1999): 209–35.

Ambrosoli, Mauro. *The Wild and the Sown: Botany and Agriculture in Western Europe, 1350–1850.* New York: Cambridge University Press, 1997.

Anderson, Abraham. *The Treatise of the Three Impostors and the Problem of Enlightenment.* Lanham, MD: Rowman & Littlefield, 1997.

Anderson, M. S., Kay J. Anderson, and Robert Anderson. *War and Society in Europe of the Old Regime 1618–1789.* Montreal: McGill–Queens University Press, 1998.

Barriendos, Mariano. "Climatic Variations in the Iberian Peninsula during the Late Maunder Minimum (AD 1675–1715): An Analysis of Data from Rogation Ceremonies." *The Holocene* 7 (1997): 105–11.

Bartuschat, Wolfgang. *Baruch de Spinoza.* Munich: C. H. Beck, 2006.

Bauer, Oswald. *Zeitungen vor der Zeitung: Die Fuggerzeitungen (1568–1605) und das frühmoderne Nachrichtensystem.* Munich: Oldenbourg Verlag, 2011.

Bayle, Pierre. *Dictionnaire historique et critique.* 2 vols. Rotterdam: Reinier Leers, 1697.

———. *Pensées diverses, écrites à un docteur de Sorbonne, à l'occasion de la comète qui parut au mois de décembre 1680*. Rotterdam: Reinier Leers, 1683–1694.

———. *Various Thoughts on the Occasion of a Comet*, trans. Robert C. Bartlett. Albany: State University of New York Press, 2000.

Beaud, Michel. *Histoire du capitalisme: 1500–2010*. 6th ed. Paris: Points, 2010.

Beer, Barrett L. *Rebellion and Riot: Popular Disorder in England During the Reign of Edward VI*. Kent, OH: Kent State University Press, 2005.

Behringer, Wolfgang. *A Cultural History of Climate*. Cambridge, UK, and Malden, MA: Polity Press, 2009.

Behringer, Wolfgang, and J. C. Grayson. *Witchcraft Persecutions in Bavaria: Popular Magic, Religious Zealotry and Reason of State in Early Modern Europe*. Cambridge: Cambridge University Press, 2003.

Behringer, Wolfgang, Hartmut Lehmann, and Christian Pfister. *Cultural Consequences of the "Little Ice Age."* Göttingen, Germany: Vandenhoeck & Ruprecht, 2005.

Beik, William. *Urban Protest in Seventeenth-Century France: The Culture of Retribution*. Cambridge: Cambridge University Press, 1997.

Benedict, Philip, and Myron P. Gutmann. *Early Modern Europe: From Crisis to Stability*. Newark: University of Delaware Press, 2005.

Benitez, Miguel. *La Face Cachée des Lumières: Recherches sur les Manuscrits Philosophiques Clandestins de l'Âge Classique*. Paris and Oxford: Voltaire Foundation, 1996.

Berman, David. *A History of Atheism in Britain: From Hobbes to Russell*. Reprint. Abingdon, UK: Routledge, 1990.

Bitton, Davis. *The French Nobility in Crisis, 1560–1640*. Stanford, CA: Stanford University Press, 1969.

Blaak, Jeroen. *Literacy in Everyday Life: Reading and Writing in Early Modern Dutch Diaries*. Leiden: Brill, 2009.

Black, Jeremy. *The Atlantic Slave Trade in World History*. Abingdon, UK: Routledge, 2015.

———. *Beyond the Military Revolution: War in the Seventeenth-Century World*. London: Palgrave Macmillan, 2011.

———. *European Warfare, 1494–1660*. London: Psychology Press, 2002.

Bletschacher, Richard. *Die Lauten- und Geigenmacher des Füssener Landes*. Hofheim am Taunus, Germany: F. Hofmeister, 1978.

Blom, Philipp. *A Wicked Company: The Forgotten Radicalism of the European Enlightenment*. New York: Basic Books, 2012.

Bödiker, Johann. *Christlicher Bericht von Cometen, als der große Comet 1680 und 1681 geleuchtet*, Köln an der Spree, Germany: Georg Schulke, 1681.

Bohanan, Donna. *Crown and Nobility in Early Modern France*. London: Palgrave Macmillan, 2001.

Bohstedt, Professor John. *The Politics of Provisions: Food Riots, Moral Economy, and Market Transition in England, c. 1550–1850*. Farnham, UK: Ashgate Publishing, 2013.

Bois, Guy. *Crise du féodalisme*. Paris: Presses de Sciences Po, 1976.

Boner, Patrick, ed. *Change and Continuity in Early Modern Cosmology*. Dordrecht, Germany: Springer Science & Business Media, 2011.

Bowen, James. *A History of Western Education. Vol. 1, The Ancient World: Orient and Mediterranean 2000 B.C.–A.D. 1054*. Abingdon, UK: Routledge, 2003.

———. *A History of Western Education. Vol. 2, Civilization of Europe, Sixth to Sixteenth Century*. New York: St. Martin's Press, 1975.

———. *A History of Western Education. Vol. 3, The Modern West: Europe and the New World*. New York: St. Martin's Press, 1981.

Bradley, Raymond S., and Philip D. Jones. *Climate since AD 1500*. Abingdon, UK: Routledge, 2003.

Brakensiek, Stefan. *Fürstendiener, Staatsbeamte, Bürger: Amtsführung und Lebenswelt der Ortsbeamten in niederhessischen Kleinstädten (1750–1830)*. Göttingen, Germany: Vandenhoeck & Ruprecht, 1999.

Braudel, Fernand. *Civilization and Capitalism, 15th–18th Century. Vol. 3, The Perspective of the World*. Oakland: University of California Press, 1992.

———. *Civilization and Capitalism, 15th–18th Century. Vol. 1, The Structures of Everyday Life*. Oakland: University of California Press, 1992.

———. *Civilization and Capitalism, 15th–18th Century. Vol. 2, The Wheels of Commerce*. Oakland: University of California Press, 1992.

Brázdil, R., O. Kotyza, P. Dobrovolný, L. Řezníčková, and H. Valášek. *Climate of the Sixteenth Century in the Czech Lands*. Brno, Czech Republic: Masaryk University, 2013.

Briffa, K. R., P. D. Jones, R. B. Vogel, F. H. Schweingruber, M. G. L. Baillie, S. G. Shiyatov, and E. A. Vaganov. "European Tree Rings and Climate in the 16th Century." *Climatic Change* 43 (1999): 151–68.

Brooke, John L. *Climate Change and the Course of Global History: A Rough Journey*. Cambridge: Cambridge University Press, 2014.

Brückner, Eduard. *Eduard Brückner: The Sources and Consequences of Climate Change and Climate Variability in Historical Times*, edited by Nico Stehr and H. von Storch. Dordrecht, Germany: Kluwer Academic Publishers, 2000.

Buc, Philippe. *Heiliger Krieg: Gewalt im Namen des Christentums*. Darmstadt, Germany: Verlag Philipp von Zabern in Wissenschaftliche Buchgesellschaft, 2015.

Buckley, Veronica. *Christina, Queen of Sweden: The Restless Life of a European Eccentric*. New York: Harper Perennial, 2005.

———. *The Secret Wife of Louis XIV: Françoise d'Aubigné, Madame de Maintenon*. Reprint. New York: Picador, 2010.

Büntgen, Ulf, and Lena Hellmann. "The Little Ice Age in Scientific Perspective: Cold Spells and Caveats." *Journal of Interdisciplinary History* 44 (2013): 353–68.

Buisman, J., and A. F. V. van Engelen. *Duizend jaar weer, wind en water in de Lage Landen*. Franeker, Netherlands: Van Wijnen, 1995.

Buisseret, David. *Sully and the Growth of Centralized Government in France: 1598–1610*. London: Eyre & Spottiswoode, 1968.

Burg, David F. *A World History of Tax Rebellions: An Encyclopedia of Tax Rebels, Revolts, and Riots from Antiquity to the Present*. Abingdon, UK: Routledge, 2004.

Burns, William E. *Science in the Enlightenment: An Encyclopedia*. Santa Barbara, CA: ABC-CLIO, 2003.

―――. *The Scientific Revolution: An Encyclopedia*. Santa Barbara, CA: ABC-CLIO, 2001.

Bury, J. B. *A History of Freedom of Thought*. Teddington, England: Echo Library, 2006.

Busche, Hubertus, ed. *Departure for Modern Europe: A Handbook of Early Modern Philosophy (1400–1700)*. Hamburg: Felix Meiner Verlag, 2011.

Buys, Ruben. *De kunst van het weldenken: Lekenfilosofie en volkstalig rationalisme in de Nederlanden (1550–1600)*. Amsterdam: Amsterdam University Press, 2009.

Byrne, James M. *Religion and the Enlightenment from Descartes to Kant*. Louisville, KY: Westminster John Knox Press, 1997.

Campbell, SueEllen. *The Face of the Earth: Natural Landscapes, Science, and Culture*. Oakland: University of California Press, 2011.

Campbell, Thomas J. *The Jesuits, 1534–1921: A History of the Society of Jesus from Its Foundation to the Present Time*. Lulu Press, 2014.

Camuffo, Dario, Chiara Bertolin, Patrizia Schenal, Alberto Craievich, and Rossella Granziero. "The Little Ice Age in Italy from Documentary Proxies and Early Instrumental Records." *Méditerranée* 122 (2014): 17–30.

Čechura, Jaroslav. *Adelige Grundherrn als Unternehmer: Zur Struktur südböhmischer Dominien vor 1620*. Munich: Oldenbourg, 2000.

Chang, Chun-shu, and Shelley Hsueh-lun Chang. *Crisis and Transformation in Seventeenth-Century China: Society, Culture, and Modernity in Li Yu's World*. Ann Arbor: University of Michigan Press, 1998.

Christian, William A. *Local Religion in Sixteenth-Century Spain*. Princeton, NJ: Princeton University Press, 1989.

Clark, Henry C. *Compass of Society: Commerce and Absolutism in Old-Regime France*. Lanham, MD: Rowman & Littlefield (Lexington Books), 2007.

Clark, Stuart. *Thinking with Demons: The Idea of Witchcraft in Early Modern Europe*. Oxford: Oxford University Press, 1999.

Cleaveland, Malcolm, David Stahle, Matthew Therrell, José Villanueva-Diaz, and Barney Burns. "Tree-Ring Reconstructed Winter Precipitation and Tropical Teleconnections in Durango, Mexico." *Climatic Change* 59 (2003): 369–88.

Clulee, Nicholas. *John Dee's Natural Philosophy: Between Science and Religion*. Abingdon, UK: Routledge, 2013.

Cohn, Norman. *The Pursuit of the Millennium: Revolutionary Millenarians and Mystical Anarchists of the Middle Ages*. New York: Random House, 1983.

Coler, Johann. *Oeconomia Ruralis Et Domestica: Darin[n] das gantz Ampt aller trewer Hauß-Väter/Hauß-Mütter/beständiges und allgemeines Hauß-Buch/vom Haußhalten/Wein- Acker- Gärten- Blumen und Feld-Baw/begriffe . . . ; Sampt beygefügter einer experimentalischer Hauß-Apotecken und kurtzer Wundartzney-Kunst/ wie dann auch eines Calendarii perpetui. . . .* Heyll, 1665.

Collins, Randall. *The Sociology of Philosophies*. Cambridge, MA: Harvard University Press, 2009.

Comenius, Johann Amos. *Joh. Amos Comenii Orbis Sensualium Pictus: Hoc Est Omnium Principalium in Mundo Rerum, & in Vita Actionum, Pictura & Nomenclatura* [including English translation]. S. Leacroft, 1777.

Condren, Conal, Stephen Gaukroger, and Ian Hunter, eds. *The Philosopher in Early Modern Europe: The Nature of a Contested Identity.* Cambridge: Cambridge University Press, 2006.

Cook, Harold John, and Sven Dupré. *Translating Knowledge in the Early Modern Low Countries.* Münster, Germany: LIT Verlag, 2012.

Coudert, Allison. *Religion, Magic, and Science in Early Modern Europe and America.* Santa Barbara, CA: ABC-CLIO, 2011.

Cressy, David. "Literacy in Seventeenth-Century England: More Evidence." *The Journal of Interdisciplinary History* 8, no. 1 (1977): 141–50.

Crosby, Alfred W., Jr. *The Columbian Exchange: Biological and Cultural Consequences of 1492, 30th Anniversary ed.* Westport, CT: Greenwood, 2003.

Cullen, Karen J. *Famine in Scotland: The "Ill Years" of the 1690s.* Scottish Historical Review monographs series, no. 16. Edinburgh: Edinburgh University Press, 2010.

Curran, Mark. *Atheism, Religion and Enlightenment in Pre-Revolutionary Europe.* Woodbridge, UK, and Rochester, NY: Royal Historical Society and Boydell Press, 2012.

Curtin, Philip D. *The Atlantic Slave Trade: A Census.* Madison: University of Wisconsin Press, 1972.

D'Addario, Christopher. *Exile and Journey in Seventeenth-Century Literature.* Cambridge: Cambridge University Press, 2007.

Darnton, Robert. *The Literary Underground of the Old Regime.* Reprint. Cambridge, MA: Harvard University Press, 1985.

"Das Leben eines Söldners." http://mvonmueller.de/Magisterarbeit_MvM_01_02_2005.pdf. Accessed February 24, 2016.

Davi, Nicole, Rosanne D'Arrigo, Gordon Jacoby, Brendan Buckley, and Osamu Kobayashi. "Warm-Season Annual to Decadal Temperature Variability for Hokkaido, Japan, Inferred from Maximum Latewood Density (AD 1557–1990) and Ring Width Data (AD 1532–1990)." *Climatic Change* 52 (2002): 201–17.

Davies, K. G. *The North Atlantic World in the Seventeenth Century.* Minneapolis: University of Minnesota Press, 1974.

Dear, Peter. *Revolutionizing the Sciences: European Knowledge and Its Ambitions, 1500–1700.* 2nd edition. Princeton, NJ: Princeton University Press, 2009.

Debus, A. G. *Science and Education in the Seventeenth Century.* London: Macdonald & Co., 1967.

Dee, John. *The Private Diary of Dr. John Dee: And the Catalogue of His Library of Manuscripts, from the Original Manuscripts in the Ashmolean Museum at Oxford, and Trinity College Library, Cambridge.* London: Camden Society, 1842.

Dee, John, and Gerald Suster. *John Dee: Essential Readings.* Berkeley, CA: North Atlantic Books, 2003.

Descartes, René. *Discours de la méthode pour bien conduire sa raison et chercher la vérité dans les sciences.* Leiden: Ian Maire, 1637.

———. *Meditationes de prima philosophia.* Paris: Michael Soly, 1641.

de Vries, Jan. "The Crisis of the Seventeenth Century: The Little Ice Age and the Mystery of the 'Great Divergence.'" *Journal of Interdisciplinary History* 44 (2013): 369–77.

———. *The Economy of Europe in an Age of Crisis, 1600–1750*. Cambridge, UK, and New York: Cambridge University Press, 1976.

Dezileau, L., P. Sabatier, P. Blanchemanche, B. Joly, D. Swingedouw, C. Cassou, J. Castaings, P. Martinez, and U. Von Grafenstein. "Intense storm activity during the Little Ice Age on the French Mediterranean coast." *Palaeogeography, Palaeoclimatology, Palaeoecology* 299, no. 1–2 (January 2011): 289–97.

Diamond, Jared M. *Guns, Germs, and Steel: The Fates of Human Societies.* New York: W. W. Norton, 1997.

Downing, Brian M. *The Military Revolution and Political Change: Origins of Democracy and Autocracy in Early Modern Europe*. Princeton, NJ: Princeton University Press, 1993.

Dull, Robert A., Richard J. Nevle, William I. Woods, Dennis K. Bird, Shiri Avnery, and William M. Denevan. "The Columbian Encounter and the Little Ice Age: Abrupt Land Use Change, Fire, and Greenhouse Forcing." *Annals of the Association of American Geographers* 100, no. 4 (2010): 755–71.

Dülmen, Richard van. *Entstehung des frühneuzeitlichen Europa 1550–1648*. Illustrated edition. Fischer World History. Frankfurt am Main: Fischer Taschenbuch Verlag, 1982.

———. "Kultur und Alltag in der frühen Neuzeit. 3. Religion, Magie, Aufklärung: 16.–18. Jahrhundert." Beck, 1994.

———. *Kultur und Alltag in der frühen Neuzeit: Dorf und Stadt: 16.–18. Jahrhundert*. C. H. Beck, 2005.

Dülmen, Richard van, Sina Rauschenbach, and Meinrad von Engelberg, eds. *Macht des Wissens: die Entstehung der modernen Wissensgesellschaft*. Köln: Böhlau, 2004.

Dupré, Louis K. *The Enlightenment and the Intellectual Foundations of Modern Culture*. New Haven, CT: Yale University Press, 2004. http://site.ebrary.com/id/10170884.

Egmond, Florike. *The World of Carolus Clusius: Natural History in the Making, 1550–1610*. Abingdon, UK: Routledge, 2015.

Eisenstein, Elizabeth L. *The Printing Press as an Agent of Change*. Cambridge: Cambridge University Press, 1980.

———. *The Printing Revolution in Early Modern Europe*. Cambridge: Cambridge University Press, 2005.

Ekelund, Robert Burton. *Politicized Economies: Monarchy, Monopoly, and Mercantilism*. College Station: Texas A&M University Press, 1997.

Eliav-Feldon, Miriam. *Renaissance Impostors and Proofs of Identity*. London: Palgrave Macmillan, 2012.

Encyclopaedia Universalis. *Traité de l'œconomie politique d'Antoine de Montchrestien: Les Fiches de lecture d'Universalis*. Encyclopaedia Universalis, 2015.

Erikson, Emily. *Between Monopoly and Free Trade: The English East India Company, 1600–1757*. Princeton, NJ: Princeton University Press, 2014.

Evelyn, John. *The Diary of John Evelyn (Complete)*. Library of Alexandria, 2013.

Fagan, Brian M. *The Little Ice Age: How Climate Made History 1300–1850.* New York: Basic Books, 2001.

Fairchilds, Cissie C. *Women in Early Modern Europe, 1500–1700.* London: Pearson Education, 2007.

Findlen, Paula. *Athanasius Kircher: The Last Man Who Knew Everything.* Abingdon, UK: Routledge, 2004.

Finel-Honigman, Irene. *A Cultural History of Finance.* Abingdon, UK: Routledge, 2009.

Fisher, Saul. *Pierre Gassendi's Philosophy And Science: Atomism for Empiricists.* Leiden: Brill, 2005.

Fleming, John V. *The Dark Side of the Enlightenment: Wizards, Alchemists, and Spiritual Seekers in the Age of Reason.* New York: W. W. Norton, 2013.

Fletcher, Anthony, and John Stevenson. *Order and Disorder in Early Modern England.* Cambridge: Cambridge University Press, 1987.

Flynn, Dennis Owen. *World Silver and Monetary History in the 16th and 17th Centuries.* Variorum, 1996.

Forgeng, Jeffrey L. *Daily Life in Stuart England.* Westport, CT: Greenwood Publishing Group, 2007.

Fouke, Daniel C. *Philosophy and Theology in a Burlesque Mode: John Toland and "The Way of Paradox."* Amherst, NY: Humanity Books, 2007.

Frei, Walter. *Der Luzerner Stadtschreiber Renward Cysat 1545–1614.* Luzern: Commissions Verlag Haag, 1963.

French, Peter J. *John Dee: The World of the Elizabethan Magus.* Abingdon, UK: Routledge, 2013.

Gall, Lothar. *Von der ständischen zur bürgerlichen Gesellschaft.* Berlin: Walter de Gruyter, 2012.

Garnier, Emmanuel. *Les dérangements du temps: 500 ans de chaud et de froid en Europe.* Paris: Plon, 2010.

Garrett, Don. *The Cambridge Companion to Spinoza.* Cambridge: Cambridge University Press, 1996.

Gay, Peter. *The Enlightenment: The Rise of Modern Paganism.* New York: W. W. Norton, 1995.

Gebhardt, Carl. *Spinoza: Briefwechsel.* Books on Demand, 2012.

Geier, Manfred. *Aufklärung: Das europäische Projekt.* Reinbek bei Hamburg: Rowohlt, 2012.

Geneva, Ann. *Astrology and the Seventeenth Century Mind: William Lilly and the Language of the Stars.* Manchester, UK: Manchester University Press, 1995.

Glaser, Rüdiger. *Klimageschichte Mitteleuropas: 1200 Jahre Wetter, Klima, Katastrophen.* Darmstadt, Germany: Primus-Verlag, 2008.

Glaser, Rudiger, Rudolf Brazdil, Christian Pfister, and Petr Dobrovolny. "Seasonal Temperature and Precipitation Fluctuations in Selected Parts of Europe during the Sixteenth Century." *Climatic Change* 43 (1999): 169–200.

Glasgow Mechanics' Magazine, and Annals of Philosophy. W. R. M'Phun, 1826.

Glassie, John. *A Man of Misconceptions: The Life of an Eccentric in an Age of Change.* New York: Penguin, 2012.

———. *Der letzte Mann, der alles wusste: Das Leben des exzentrischen Genies Athanasius Kircher*. eBook Berlin Verlag, 2014.

Glete, Jan. *Warfare at Sea, 1500–1650: Maritime Conflicts and the Transformation of Europe*. London and New York: Routledge, 2000.

Goldgar, Anne. *Tulipmania: Money, Honor, and Knowledge in the Dutch Golden Age*. Chicago: University of Chicago Press, 2008.

Goldstone, Jack A. *Revolution and Rebellion in the Early Modern World*. Oakland: University of California Press, 1991.

Goor, Jurrien van, and Foskelien van Goor. *Prelude to Colonialism: The Dutch in Asia*. Hilversum, Netherlands: Uitgeverij Verloren, 2004.

Goubert, Pierre. *The French Peasantry in the Seventeenth Century*. Cambridge: Cambridge University Press, 1986.

Graff, Harvey J. *Literacy and Social Development in the West: A Reader*. Cambridge: CUP Archive, 1981.

———. *Understanding Literacy in Its Historical Contexts: Socio-Cultural History and the Legacy of Egil Johansson*. Lund, Sweden: Nordic Academic Press, 2009.

Grayling, A. C. *The Age of Genius: The Seventeenth Century and the Birth of the Modern Mind*. London: Bloomsbury Publishing, 2016.

Greengrass, Mark. *Christendom Destroyed: Europe 1517–1648*. London: Allen Lane, 2014.

Grigg, D. B. *Population Growth and Agrarian Change: An Historical Perspective*. Cambridge: CUP Archive, 1980.

Grotius, Hugo. *Annales et histoires des troubles du Pays-Bas, par Hugo Grotius*. Amsterdam: L'imprimerie de Jean Blaeu, 1662.

———. *Annales et Historiae de Rebus Belgicis*. http://archive.org/details/annalesethistor00grotgoog. Accessed April 12, 2016.

———. *Hug. Grotii Poema ta omnia*. Apud Hieronymum de Vogel, 1645.

Grove, Jean M. *Little Ice Ages: Ancient and Modern*. Abingdon, UK: Routledge, 2013.

———. *The Little Ice Age*. Abingdon, UK: Routledge, 2012.

Grunert, Frank. *Concepts of (radical) Enlightenment: Jonathan Israel in Discussion*. Halle (Saale): Mitteldeutscher Verlag, 2014.

Guiot, Joel, and Christophe Corona. "Growing Season Temperatures in Europe and Climate Forcings Over the Past 1400 Years." *PLOS ONE* 5, no. 4 (April 2010).

Hale, John Rigby. *War and Society in Renaissance Europe, 1450–1620*. Montreal: McGill-Queen's University Press, 1998.

Hall, David D. *Cultures of Print: Essays in the History of the Book*. Amherst: University of Massachusetts Press, 1996.

Harkness, Deborah E. *John Dee's Conversations with Angels: Cabala, Alchemy, and the End of Nature*. Cambridge: Cambridge University Press, 1999.

Harris, Jonathan Gil. *Sick Economies: Drama, Mercantilism, and Disease in Shakespeare's England*. Philadelphia: University of Pennsylvania Press, 2013.

Hazard, Paul, and Anthony Grafton. *The Crisis of the European Mind: 1680–1715*. Translated by J. Lewis May. New York: NYRB Classics, 2013.

He, Minhui, Bao Yang, Achim Bräuning, Jianglin Wang, and Zhangyong Wang. "Tree-Ring Derived Millennial Precipitation Record for the South-Central

Tibetan Plateau and Its Possible Driving Mechanism." *The Holocene* 23, no. 1 (2013): 36–45.

Heal, Felicity, and Clive Holmes. *The Gentry in England and Wales, 1500–1700.* Stanford, CA: Stanford University Press, 1994.

Heckscher, Eli F. *Mercantilism.* Abingdon, UK: Routledge, 2013.

Heinrich, Ingo, Ramzi Touchan, Isabel Dorado Liñán, Heinz Vos, and Gerhard Helle. "Winter-to-Spring Temperature Dynamics in Turkey Derived from Tree Rings since AD 1125." *Climate Dynamics* 41 (2013): 1685–1701.

Hennig, I. R. *Katalog bemerkenswerter Witterereignisse von den ältesten Zeiten bis zum Jahre 1800.* Berlin: Royal Prussian Academy, 1904.

Hersche, Peter. *Muße und Verschwendung: Europäische Gesellschaft und Kultur im Barockzeitalter.* Verlag Herder GmbH, 2015.

Hetherington, Norriss S. "Almanacs and the Extent of Knowledge of the New Astronomy in Seventeenth-Century England." *Proceedings of the American Philosophical Society* 119, no. 4 (1975): 275–79.

Holt, Mack P. "Review of *Fiscal Limits of Absolutism: Direct Taxation in Early Seventeenth-Century France*, by James B. Collins." *Journal of Economic History* 50, no. 1 (2009): 191–92.

Hopcroft, Rosemary Lynn. *Regions, Institutions, and Agrarian Change in European History.* Ann Arbor: University of Michigan Press, 1999.

Houston, Rab. *Literacy in Early Modern Europe.* Abingdon, UK: Routledge, 2014.

Hoyle, Richard W., ed. *Custom, Improvement and the Landscape in Early Modern Britain.* Farnham, UK: Ashgate Publishing, 2011.

Hudson, Wayne. *The English Deists: Studies in Early Enlightenment.* London and Brookfield, VT: Pickering & Chatto, 2009.

Huff, Toby E. *Intellectual Curiosity and the Scientific Revolution: A Global Perspective.* Cambridge: Cambridge University Press, 2010.

Hughes, Philip D. "Little Ice Age Glaciers in the Mediterranean Mountains." *Méditerranée* 122 (2015): 63–79.

Hunger, F. W. T. *Charles de l'Escluse (Carolus Clusius) Nederlandsch kruidkundige, 1526–1609.* The Hague: Martinus Nijhoff, 1927.

Irvin, Jeffery L., ed. *Paradigm and Praxis: Seventeenth-Century Mercantilism and the Age of Liberalism.* Toledo, OH: University of Toledo ProQuest, 2008.

Israel, Jonathan Irvine. *European Jewry in the Age of Mercantilism, 1550–1750.* Liverpool: Liverpool University Press, Littman Library of Jewish Civilization, 1998.

———. *Radical Enlightenment: Philosophy and the Making of Modernity, 1650–1750.* Oxford and New York: Oxford University Press, 2001.

———. *A Revolution of the Mind: Radical Enlightenment and the Intellectual Origins of Modern Democracy.* Princeton, NJ: Princeton University Press, 2011.

———. *Revolutionary Ideas: An Intellectual History of the French Revolution from The Rights of Man to Robespierre.* Princeton, NJ: Princeton University Press, 2014.

Jacob, Margaret C. *The Radical Enlightenment: Pantheists, Freemasons, and Republicans.* London and Boston: Allen & Unwin, 1981.

Jäger, Georg. "Hochweidewirtschaft, Klimaverschlechterung ('Kleine Eiszeit') und Gletschervorstöße in Tirol zwischen 1600 und 1850," in *Tiroler Hei-mat, Jahrbuch für Geschichte und Volkskunde* 70 (2006): 5–83.

Jakubowski-Tiessen, Manfred. *Jahrhundertwenden: Endzeit- und Zukunftsvorstellun-gen vom 15. bis zum 20. Jahrhundert.* Göttingen, Germany: Vandenhoeck & Ruprecht, 1999.

———, ed. *Krisen des 17. Jahrhunderts: Interdisziplinäre Perspektiven.* Sammlung Van-denhoeck. Göttingen, Germany: Vandenhoeck & Ruprecht, 1999.

James, Lawrence. *The Middle Class: A History.* London: Hachette UK, 2010.

Jaser, Christian, Heinz Schilling, and Stefan Ehrenpreis. *Erziehung und Schulwe-sen zwischen Konfessionalisierung und Säkularisierung.* Münster, Germany: Wax-mann Verlag, 1963.

Jin, Dengjian. *The Great Knowledge Transcendence: The Rise of Western Science and Tech-nology Reframed.* London: Palgrave Macmillan, 2015.

Joas, Hans, and Wolfgang Knöbl. *War in Social Thought: Hobbes to the Present.* Princeton, NJ: Princeton University Press, 2013.

Johnson, Carina L. *Cultural Hierarchy in Sixteenth-Century Europe: The Ottomans and Mexicans.* New York: Cambridge University Press, 2011.

Jonkanski, Dirk. "Oberdeutsche Baumeister in Venedig. Reiserouten und Besichtigungsprogramme," Mainz: Akademie der Wissenschaft und der Literatur, 1993.

Jue, Jeffrey K. *Heaven Upon Earth: Joseph Mede (1586–1638) and the Legacy of Millena-rianism.* Berlin: Springer Science & Business Media, 2006.

Jutikkala, Eino. "The Great Finnish Famine in 1696–97." *Scandinavian Economic History Review* 3 (1955): 48–63.

Kamen, Henry. *Early Modern European Society.* London: Psychology Press, 2000.

Kaniewski, David, and Élise Van Campo. "Pollen-Inferred Palaeoclimatic Pat-terns in Syria during the Little Ice Age." *Méditerranée* 122 (2015): 139–44.

Kaufmann, Thomas. *Konfession und Kultur: Lutherischer Protestantismus in der zweiten Hälfte des Reformationsjahrhunderts.* Mohr Siebeck, 2006.

Kelly, Morgan, and Cormac Ó Gráda. "Change Points and Temporal Depen-dence in Reconstructions of Annual Temperature: Did Europe Experi-ence a Little Ice Age?" *The Annals of Applied Statistics* 8, no. 3 (September 2014): 1372–94.

Kerridge, Eric. *The Agricultural Revolution.* Abingdon, UK: Routledge, 2013.

Kettering, Sharon. *Judicial Politics and Urban Revolt in Seventeenth-Century France: The Parlement of Aix, 1629–1659.* Princeton, NJ: Princeton University Press, 2015.

Kindleberger, Charles Poor. *Historical Economics: Art or Science?* Oakland: Univer-sity of California Press, 1990.

Kiple, Kenneth F., and Kriemhild Coneè Ornelas, eds. *The Cambridge World His-tory of Food.* Cambridge: Cambridge University Press, 2000.

Knoll, Martin, and Reinhold Reith. *An Environmental History of the Early Modern Period: Experiments and Perspectives.* Münster: LIT Verlag, 2014.

Knox, MacGregor, and Williamson Murray. *The Dynamics of Military Revolution, 1300–2050.* Cambridge: Cambridge University Press, 2001.

Koenigsberger, H. G. *Early Modern Europe 1500–1789.* Abingdon, UK: Routledge, 2014.

Koenigsberger, H. G., George L. Mosse, and G. Q. Bowler. *Europe in the Sixteenth Century.* Routledge, 2014.

Koerbagh, Johannes, and Adriaan Koerbagh. *Een ligt schijnende in duystere plaatsen.* Vlaamse Vereniging voor Wijsbegeerte, 1974.

Köse, N., Ü. Akkemik, H. N. Dalfes, and M. S. Özeren. "Tree-Ring Reconstructions of May–June Precipitation for Western Anatolia." *Quaternary Research* 75 (2011): 438–50.

Koslowski, Gerhard, and Rüdiger Glaser. "Variations in Reconstructed Ice Winter Severity in the Western Baltic from 1501 to 1995, and Their Implications for the North Atlantic Oscillation." *Climatic Change* 41 (February 1999): 175–91.

Kottman, Karl A. *Millenarianism and Messianism in Early Modern European Culture. Volume II: Catholic Millenarianism: From Savonarola to the Abbé Grégoire.* Berlin: Springer Science & Business Media, 2013.

Kwass, Michael. *Privilege and the Politics of Taxation in Eighteenth-Century France: Liberté, Egalité, Fiscalité.* Cambridge: Cambridge University Press, 2006.

L'Estoile, Pierre de. *Mémoires-journaux de Pierre de L'Estoile.* Paris: Tallandier, 1875.

Lach, Donald F., and Edwin J. Van Kley. *Asia in the Making of Europe. Volume III: A Century of Advance. Book 1: Trade, Missions, Literature.* Chicago: University of Chicago Press, 1998.

Lachiver, Marcel. *Les années de misère: La famine au temps du Grand Roi, 1680–1720.* Paris: Fayard, 1991.

Ladurie, Emmanuel Le Roy. *The French Peasantry, 1450–1660.* Oakland: University of California Press, 1987.

———. *Histoire du climat depuis l'an mil.* Paris: Flammarion, 1967.

Laerke, Mogens. *The Use of Censorship in the Enlightenment.* Leiden: Brill, 2009.

Laffemas, Barthélemy de (1545–1612?). *Les discours d'une liberté générale et vie heureuse pour le bien du peuple. Composé par Barthélemy de Laffemas . . . ,* 1601.

Lafontaine-Boyer, Karelle, and Konrad Gajewski. "Vegetation Dynamics in Relation to Late Holocene Climate Variability and Disturbance, Outaouais, Québec, Canada." *The Holocene* 24, no. 11 (November 2014): 1515–26.

Lamb, Hubert H. *Climate, History and the Modern World.* Abingdon, UK: Routledge, 2002.

Landsteiner, Erich. "The Crisis of Wine Production in Late Sixteenth-Century Central Europe: Climatic Causes and Economic Consequences." *Climatic Change* 43, no. 1 (September 1999): 323–34.

Landwehr, Achim. *Geburt der Gegenwart: Eine Geschichte der Zeit im 17. Jahrhundert.* Frankfurt: S. Fischer Verlag, 2014.

Laursen, John Christian, and R. H. Popkin. *Millenarianism and Messianism in Early Modern European Culture. Volume IV: Continental Millenarians: Protestants, Catholics, Heretics.* Berlin: Springer Science & Business Media, 2013.

Law, Rev. Robert. *Memorialls; or The Memorable Things That Fell Out Within This Island of Britain from 1638 to 1684* (Edinburgh: Archibald Constable and Co., 1818).

Law, Robin, Suzanne Schwarz, and Silke Strickrodt, eds. *Commercial Agriculture, the Slave Trade and Slavery in Atlantic Africa.* Martelsham, UK: Boydell & Brewer, 2013.

Levack, Brian P. *Demonology, Religion, and Witchcraft: New Perspectives on Witchcraft, Magic, and Demonology.* Abingdon, UK: Routledge, 2013.

———. *Hexenjagd: Die Geschichte der Hexenverfolgung in Europa.* Munich: C. H. Beck, 2009.

Li, Zong-Shan, Qi-Bin Zhang, and Keping Ma. "Tree-ring Reconstruction of Summer Temperature for A.D. 1475–2003 in the Central Hengduan Mountains, Northwestern Yunnan, China." *Climatic Change* 110 (2012): 455–67.

Lindberg, David C. *The Beginnings of Western Science: The European Scientific Tradition in Philosophical, Religious, and Institutional Context, Prehistory to A.D. 1450.* 2nd edition. Chicago: University of Chicago Press, 2008.

Liu, Jingjing, Bao Yang, and Chun Qin. "Tree-Ring Based Annual Precipitation Reconstruction since AD 1480 in South Central Tibet." *Quaternary International* 236 (2011): 75–81.

Lloyd, Genevieve. *Enlightenment Shadows.* Oxford: Oxford University Press, 2013.

LoLordo, Antonia. *Pierre Gassendi and the Birth of Early Modern Philosophy.* Cambridge: Cambridge University Press, 2006.

Löwe, Heinz-Dietrich. *Volksaufstände in Russland: Von der Zeit der Wirren bis zur "Grünen Revolution" gegen die Sowjetherrschaft.* Wiesbaden: Otto Harrassowitz Verlag, 2006.

Luckman, B. H., K. R. Briffa, P. D. Jones, and F. H. Schweingruber. "Tree-Ring Based Reconstruction of Summer Temperatures at the Columbia Icefield, Alberta, Canada, AD 1073–1983." *The Holocene* 7 (1997): 375–89.

Lund, David C., Jean Lynch-Stieglitz, and William B. Curry. "Gulf Stream Density Structure and Transport during the Past Millennium." *Nature* 444 (2006): 601–4.

Lundgreen, Peter. *Sozial- und Kulturgeschichte des Bürgertums: Eine Bilanz des Bielefelder Sonderforschungsbereichs (1986–1997).* Göttingen, Germany: Vandenhoeck & Ruprecht, 2000.

Lüsebrink, Hans-Jürgen. *Das Europa der Aufklärung und die aussereuropäische koloniale Welt.* Göttingen, Germany: Wallstein Verlag, 2006.

Luterbacher, Jürg, Daniel Dietrich, Elena Xoplaki, Martin Grosjean, and Heinz Wanner. "European Seasonal and Annual Temperature Variability, Trends, and Extremes Since 1500." *Science* 303 (2004): 1499–1503.

Macadam, Joyce. "English Weather: The Seventeenth-Century Diary of Ralph Josselin." *Journal of Interdisciplinary History* 43, no. 2 (2012): 221–46.

Macdonald, Alan R., and John McCallum. "The Evidence for Early Seventeenth-Century Climate from Scottish Ecclesiastical Records." *Environment and History* 19, no. 4 (November 1, 2013): 487–509.

Magagna, Victor V. *Communities of Grain: Rural Rebellion in Comparative Perspective.* Ithaca, NY: Cornell University Press, 1991.

Magnusson, Lars. *The Political Economy of Mercantilism.* Abingdon, UK: Routledge, 2015.

Mandeville, Bernard. *The Fable of the Bees; or Private Vices, Publick Benefits*, edited by Irwin Primer. New York: Capricorn Books, 1962. http://archive.org /details/fableofthebeesor027890mbp.

Mann, Michael E., Jose D. Fuentes, and Scott Rutherford. "Underestimation of Volcanic Cooling in Tree-Ring-Based Reconstructions of Hemispheric Temperatures." *Nature Geoscience* (advance online publication) (February 2012).

Mauelshagen, Franz. *Klimageschichte der Neuzeit: 1500—1900*. Darmstadt: Wiss-Buchges, 2010.

Maur, Eduard. "Gutsherrschaft und 'zweite Leibeigenschaft'" in *Böhmen: Studien zur Wirtschafts-, Sozial- und Bevölkerungsgeschichte (14.–18. Jahrhundert)*. Munich and Vienna: Verlag für Geschichte und Politik, 2001.

Mauthner, Fritz. *Spinoza: Ein Umriss seines Lebens und Wirkens (Vollständige Biografie): Baruch de Spinoza—Lebensgeschichte, Philosophie und Theologie*. e-artnow, 2015.

McCullough, Roy L. *Coercion, Conversion and Counterinsurgency in Louis XIV's France*. Leiden: Brill, 2007.

McKenna, Antony, Pierre-François Moreau, and Institut Claude Longeon. *Libertinage et philosophie au XVIIe siècle: "Gassendi et les gassendistes" et "Les passions libertines."* Saint-Étienne, France: Université de Saint-Étienne, 2000.

McManners, John. *Death and the Enlightenment: Changing Attitudes to Death among Christians and Unbelievers in Eighteenth-Century France*. Oxford: Clarendon Press, 1981.

McNeill, J. R., and Erin Stewart Mauldin. *A Companion to Global Environmental History*. New York: John Wiley, 2012.

McNeill, William Hardy. *Plagues and Peoples*. Garden City, NY: Anchor Press, 1976.

Melton, James Van Horn. *The Rise of the Public in Enlightenment Europe*. Cambridge: Cambridge University Press, 2001.

Melzer, Arthur M. *Philosophy Between the Lines: The Lost History of Esoteric Writing*. Chicago and London: University of Chicago Press, 2014.

Mengzi, and Bryan W. Van Norden. *The Essential Mengzi*. Indianapolis and Cambridge: Hackett Publishing, 2009.

Merrill, Brian L. *Athanasius Kircher (1602–1680), Jesuit Scholar: An Exhibition of His Works in the Harold B. Lee Library Collections at Brigham Young University*. Mansfield, CT: Martino Publishing, 2003.

Midelfort, H. C. Erik. *Witch Hunting in Southwestern Germany, 1562–1684: The Social and Intellectual Foundations*. Stanford, CA: Stanford University Press, 1972.

Mikami, Takehiko. "The Little Ice Age in Japan. Summer Temperature Variabilities in Japan Reconstructed from Diary Weather Records during the Little Ice Age." *Journal of Geography (Chigaku Zasshi)* 102, no. 2 (1993): 144–51.

Mikhail, Alan. *Nature and Empire in Ottoman Egypt: An Environmental History*. Cambridge and New York: Cambridge University Press, 2013.

Miller, Christopher L. *The French Atlantic Triangle: Literature and Culture of the Slave Trade*. Durham, NC: Duke University Press, 2008.

Mishra, Pankaj. *Age of Anger: A History of the Present*. London: Allen Lane, 2017.

Miskimin, Harry A. *The Economy of Later Renaissance Europe 1460–1600*. Cambridge: CUP Archive, 1977.

Modelski, George, and William R. Thompson. *Seapower in Global Politics, 1494–1993*. London: Palgrave Macmillan, 1988.

Mokyr, Joel. *The Enlightened Economy: An Economic History of Britain 1700–1850*. New Haven, CT: Yale University Press, 2010.

Montaigne, Michel de. *Essays*, trans. Charles Cotton, ed. William C. Hazlitt. London, 1877.

Moon, David. *The Russian Peasantry 1600–1930: The World the Peasants Made*. Abingdon, UK: Routledge, 2014.

Morris, T. A. *Europe and England in the Sixteenth Century*. Abingdon, UK: Routledge, 2002.

Motley, Mark Edward. *Becoming a French Aristocrat: The Education of the Court Nobility, 1580–1715*. Princeton, NJ: Princeton University Press, 2014.

Moulton, Ian Frederick, ed. *Reading and Literacy in the Middle Ages and Renaissance*. Turnhout, Belgium: Brepols, 2004.

Možný, Martin, Rudolf Brázdil, Petr Dobrovolný, and Mirek Trnka. "Cereal Harvest Dates in the Czech Republic between 1501 and 2008 as a Proxy for March–June Temperature Reconstruction." *Climatic Change* 110 (2012): 801–21.

Mrgic, Jelena. "Wine or Raki: The Interplay of Climate and Society in Early Modern Ottoman Bosnia." *Environment and History* 17 (2011): 613–37.

Muchembled, Robert, and William Monter. *Cultural Exchange in Early Modern Europe*. Cambridge: Cambridge University Press, 2006.

Muir, Edward. *Ritual in Early Modern Europe*. Cambridge: Cambridge University Press, 2005.

Mulryne, J. R. *Europa Triumphans: Court and Civic Festivals in Early Modern Europe*. Farnham, UK: Ashgate Publishing, 2005.

Mulsow, Martin. *Prekäres Wissen: Eine andere Ideengeschichte der Frühen Neuzeit*. Berlin: Suhrkamp Verlag, 2012.

Mun, Thomas. *England's Treasure by Forraign Trade*. London, 1664. Reprint, Economic History Society.

Münch, Paul. *Lebensformen in der frühen Neuzeit: 1500 bis 1800*. Neuausgabe von Ullstein-Bücher 35597. Berlin: Ullstein, 1998.

Munck, Thomas. *Seventeenth-Century Europe: State, Conflict and Social Order in Europe 1598–1700*. London: Palgrave Macmillan, 2005.

Murr, Sylvia. *Gassendi et l'Europe, 1592–1792: Actes du colloque international de Paris, "Gassendi et sa postérité (1592–1792)," Sorbonne, 6–10 octobre 1992*. Paris: Vrin, 1997.

Nadler, Steven. *A Book Forged in Hell: Spinoza's Scandalous Treatise and the Birth of the Secular Age*. Princeton, NJ: Princeton University Press, 2011 (and rev. ed., 2013).

———. *Spinoza: A Life*. Cambridge: Cambridge University Press, 2001.

———. *Spinoza's Heresy: Immortality and the Jewish Mind*. Oxford: Clarendon Press, 2001.

Nagourney, Adam, Jack Healy, and Nelson D. Schwartz. "California Drought Tests History of Endless Growth." *New York Times*, April 4, 2015.

Namer, Emile. *La vie et l'œuvre de J. C. Vanini: Prince des libertins, mort à Toulouse sur le bûcher en 1619*. Paris: Vrin, 1980.

Nazzareno, Diodato. "Climatic Fluctuations in Southern Italy since the 17th Century: Reconstruction with Precipitation Records at Benevento." *Climatic Change* 80 (2007): 411–31.

Nelson, Eric. *The Jesuits and the Monarchy: Catholic Reform and Political Authority in France* (1590–1615). Aldershot, UK, and Burlington, VT: Jesuit Historical Institute, 2005.

Nierop, Henk van. *Treason in the Northern Quarter: War, Terror, and the Rule of Law in the Dutch Revolt*. Princeton, NJ: Princeton University Press, 2009.

Nouhuys, Tabitta Van. *The Age of Two-Faced Janus: The Comets of 1577 and 1618 and the Decline of the Aristotelian World View in the Netherlands*. Leiden: Brill, 1998.

Núñez, Clara Eugenia. *Aristocracy, Patrimonial Management Strategies and Economic Development, 1450–1800*. Seville: Editorial Universidad de Sevilla, 1998.

Oltmer, Jochen. *Handbuch Staat und Migration in Deutschland seit dem 17. Jahrhundert*. Berlin: Walter de Gruyter, 2015.

O'Malley, John W., S.J., Gauvin Alexander Bailey, Steven J. Harris, and T. Frank Kennedy, S.J. *The Jesuits: Cultures, Sciences, and the Arts, 1540–1773*. Toronto: University of Toronto Press, 2006.

Overton, Mark. *Agricultural Revolution in England: The Transformation of the Agrarian Economy 1500–1850*. Cambridge: Cambridge University Press, 1996.

Paas, Martha White, John Roger Paas, and George C. Schoolfield. *The* Kipper *and* Wipper *Inflation, 1619–23: An Economic History with Contemporary German Broadsheets*. New Haven, CT: Yale University Press, 2012.

Parish, Helen, ed. *Superstition and Magic in Early Modern Europe: A Reader*. London: Bloomsbury Publishing, 2014.

Park, Katharine, and Lorraine Daston. *Cambridge History of Science: Vol. 3, Early Modern Science*. Cambridge: Cambridge University Press, 2003.

Parker, Geoffrey. "Crisis and Catastrophe: The Global Crisis of the Seventeenth Century Reconsidered." *American Historical Review* 113 (2008): 1053–79.

———. *Global Crisis: War, Climate Change and Catastrophe in the Seventeenth Century*. New Haven, CT: Yale University Press, 2013.

———. *The Military Revolution: Military Innovation and the Rise of the West, 1500–1800*. Cambridge: Cambridge University Press, 1996.

Parker, Geoffrey, and Lesley M. Smith. *The General Crisis of the Seventeenth Century*. Abingdon, UK: Routledge, 2005.

Parry, M. L. *Climate Change, Agriculture and Settlement*. Folkstone, UK: Dawson, 1978.

Pavord, Anna. *The Tulip*. New York: Bloomsbury USA, 1998.

Peacey, Jason. "Print Culture and Political Lobbying during the English Civil Wars." *Parliamentary History* 26, no. 1 (2007): 30–48.

Peakman, Julie, ed. *A Cultural History of Sexuality in the Enlightenment*. London: Bloomsbury, 2012.

Pennington, Donald. *Europe in the Seventeenth Century*. Abingdon, UK: Routledge, 2015.

Perler, Dominik. *René Descartes*. Munich: C. H. Beck, 2006.

Peters, Jan. *Peter Hagendorf—Tagebuch eines Söldners aus dem Dreißigjährigen Krieg.* Göttingen, Germany: V&R Unipress, 2012.

Pfister, Christian. *Bevölkerungsgeschichte und historische Demographie 1500–1800.* Munich: Oldenbourg Verlag, 2007.

———. *Klimageschichte der Schweiz 1525–1860.* Bern: Haupt, 1984.

Pfister, Christian, and Rudolf Brázdil. "Climatic Variability in Sixteenth-Century Europe and Its Social Dimension: A Synthesis." *Climatic Change* 43 (1999): 5–53.

Pfister, Christian, Rudolf Brázdil, and Rüdiger Glaser. *Climatic Variability in Sixteenth-Century Europe and Its Social Dimension.* Berlin: Springer Science & Business Media, 2013.

Pfister, Christian, Rudolf Brázdil, Rüdiger Glaser, and Anita Bokwa. "Daily Weather Observations in Sixteenth-Century Europe." *Climatic Change* 43 (1999): 111–50.

Pfister, Christian, Jürg Luterbacher, and Heinz Wanner. *Wetternachhersage: 500 Jahre Klimavariationen und Naturkatastrophen (1496–1995).* Bern: Haupt, 1999.

Pfister, Christian, G. Schwarz-Zanetti, and M. Wegmann. "Winter Severity in Europe: The Fourteenth Century." *Climatic Change* 34 (1996): 91–108.

Phillips, Derek L. *Well-Being in Amsterdam's Golden Age.* Amsterdam: Amsterdam University Press, 2008.

Phillips, Henry. *Church and Culture in Seventeenth-Century France.* Cambridge: Cambridge University Press, 2002.

Piervitali, E., and M. Colacino. "Evidence of Drought in Western Sicily during the Period 1565–1915 from Liturgical Offices." *Climatic Change* 49 (2001): 225–38.

Pitman, K. J., and D. J. Smith. "Tree-Ring Derived Little Ice Age Temperature Trends from the Central British Columbia Coast Mountains, Canada." *Quaternary Research* 78 (2012): 417–26.

Plummer, Marjorie Elizabeth, and Robin Barnes. *Ideas and Cultural Margins in Early Modern Germany: Essays in Honor of H.C. Erik Midelfort.* Farnham, UK: Ashgate Publishing, 2009.

Polanyi, Karl. *The Great Transformation: The Political and Economic Origins of Our Time.* 2nd ed. Boston: Beacon Press, 2001.

Porter, Roy. *The Creation of the Modern World: The Untold Story of the British Enlightenment.* New York: W. W. Norton, 2000.

———. *Flesh in the Age of Reason.* New York: W. W. Norton, 2004.

Post, John. *Food Shortage, Climatic Variability, and Epidemic Disease in Preindustrial Europe.* Ithaca, NY: Cornell University Press, 1985.

Post, John D. "Climatic Variability and the European Mortality Wave of the Early 1740s." *Journal of Interdisciplinary History* 15 (1984): 1–30.

Postma, Johannes. *The Dutch in the Atlantic Slave Trade, 1600–1815.* Cambridge: Cambridge University Press, 2008.

Potter, David. *War and Government in the French Provinces.* Cambridge: Cambridge University Press, 2003.

Prak, Maarten, and Diane Webb. *The Dutch Republic in the Seventeenth Century: The Golden Age*. Cambridge: Cambridge University Press, 2005.

Prest, Wilfrid R. *The Professions in Early Modern England*. London: Croom Helm, 1987.

Prothero, Rowland E. *English Farming, Past and Present*. Cambridge: Cambridge University Press, 2013.

Rabb, T., and R. Rotberg, eds.. *Climate and History*. Princeton, NJ: Princeton University Press, 1981.

Ravallion, Martin. *The Economics of Poverty: History, Measurement, and Policy*. Oxford: Oxford University Press, 2016.

Raven, James. *Publishing Business in Eighteenth-Century England*. Martlesham, UK: Boydell & Brewer, 2014.

Rawley, James A. *London, Metropolis of the Slave Trade*. Columbia: University of Missouri Press, 2003.

Rawley, James A., and Stephen D. Behrendt. *The Transatlantic Slave Trade: A History*. Lincoln: University of Nebraska Press, 2005.

Rebitsch, Robert. *Wallenstein: Biografie eines Machtmenschen*. Vienna: Böhlau Verlag, 2010.

Reilly, Benjamin. *Disaster and Human History: Case Studies in Nature, Society and Catastrophe*. Jefferson, NC: McFarland, 2009.

Reinders, Michel. *Printed Pandemonium: Popular Print and Politics in the Netherlands 1650–72*. Leiden: Brill, 2013.

Reith, Reinhold. *Umweltgeschichte der Frühen Neuzeit*. Berlin and Boston: Walter de Gruyter, 2011.

Richards, Jeffrey. *Sex, Dissidence and Damnation: Minority Groups in the Middle Ages*. Abingdon, UK: Routledge, 2013.

Richardson, Glenn. *The Contending Kingdoms: France and England, 1420–1700*. Farnham, UK: Ashgate Publishing, 2008.

Roberts, Benjamin. *Through the Keyhole: Dutch Child-Rearing Practices in the 17th and 18th Century: Three Urban Elite Families*. Hilversum, Netherlands: Uitgeverij Verloren, 1998.

Robinson, James Howard. *The Great Comet of 1680: A Study in the History of Rationalism*. Northfield, MN: Press of the Northfield News, 1916.

Robinson, Peter J. "Ice and Snow in Paintings of Little Ice Age Winters." *Weather* 60, no. 2 (February 2005): 37–41.

Robock, Alan. "Volcanic Eruptions and Climate." *Reviews of Geophysics* 38 (2000): 191–219.

Roeck, Bernd. *Lebenswelt und Kultur des Bürgertums in der Frühen Neuzeit*. Munich: Oldenbourg Verlag, 2011.

Rogers, Clifford J. *The Military Revolution Debate: Readings on the Military Transformation of Early Modern Europe*. Boulder, CO: Westview Press, 1995.

Rohr, Christian. *Extreme Naturereignisse im Ostalpenraum: Naturerfahrung im Spätmittelalter und am Beginn der Neuzeit*. Cologne, Germany: Böhlau, 2007.

Rösener, Werner. *Agrarwirtschaft, Agrarverfassung und ländliche Gesellschaft im Mittelalter*. Enzyklopädie deutscher Geschichte, vol. 13. Munich: Oldenbourg, 1992.

Rothkamm, Jan. *Institutio Oratoria: Bacon, Descartes, Hobbes, Spinoza*. Leiden: Brill, 2009.

Ruddiman, William. *Plows, Plagues, and Petroleum: How Humans Took Control of Climate*. Princeton, NJ: Princeton University Press, 2005.

Rüegg, Walter, and Asa Briggs. *Geschichte der Universität in Europa: Von der Reformation bis zur Französischen Revolution (1500–1800)*. Munich: C. H. Beck, 1996.

Salaman, Redcliffe N., and William Glynn Burton. *The History and Social Influence of the Potato*. Cambridge: Cambridge University Press, 1985.

Sassier, Philippe. *Du bon usage des pauvres: Histoire d'un thème politique (XVIe–XXe siècle)*. Paris: Fayard, 1990.

Sawyer, Jeffrey K. *Printed Poison: Pamphlet Propaganda, Faction Politics, and the Public Sphere in Early Seventeenth-Century France*. Oakland: University of California Press, 1990.

Schachter, Marc D. *Voluntary Servitude and the Erotics of Friendship: From Classical Antiquity to Early Modern France*. Farnham, UK: Ashgate Publishing, 2008.

Schama, Simon. *The Embarrassment of Riches: An Interpretation of Dutch Culture in the Golden Age*. New York: HarperPerennial, 1991.

Scheurleer, T. H. L., C. B. Posthumus Meyjes, and G. H. M. (Guillaume Henri Marie), eds. *Leiden University in the Seventeenth Century: An Exchange of Learning*. Leiden: Brill, 1975.

Schilling, Heinz. *Die Stadt in der frühen Neuzeit*. Munich: Oldenbourg Verlag, 2004.

Schmid, Boris V., Ulf Büntgen, W. Ryan Easterday, Christian Ginzler, Lars Walløe, Barbara Bramanti, and Nils Chr. Stenseth. "Climate-Driven Introduction of the Black Death and Successive Plague Reintroductions into Europe." *Proceedings of the National Academy of Sciences* 112 (March 2015): 3020–25.

Schneer, Jonathan. *The Thames: England's River*. London: Hachette UK, 2015.

Schnurr, Eva-Maria. *Religionskonflikt und Öffentlichkeit: Eine Mediengeschichte des Kölner Kriegs (1582 bis 1590)*. Cologne and Weimar: Böhlau Verlag, 2009.

Scholem, Gershom Gerhard, and R. J. Zwi Werblowsky. *Sabbatai Ṣevi: The Mystical Messiah, 1626–1676*. Princeton, NJ: Princeton University Press, 1973.

Schönwiese, C.-D. *Klimaschwankungen*. Heidelberg: Springer-Verlag, 2013.

Schorer, Christoph. *Memminger Chronik oder kurze Erzählung vieler denkwürdigen Sachen*. Villingen-Schwenningen, Germany: Kuhn, 1660.

Schröder-Lembke, Gertrud. *Studien zur Agrargeschichte*. Stuttgart: Lucius & Lucius, 1978.

Scott, H. M., ed. *The European Nobilities in the Seventeenth and Eighteenth Centuries*. London and New York: Longman Group UK, 1995.

Scott, Hamish. *The Oxford Handbook of Early Modern European History, 1350–1750. Vol. I: Peoples and Place*. Oxford: Oxford University Press, 2015.

———. *The Oxford Handbook of Early Modern European History, 1350–1750: Vol. II: Cultures and Power*. Oxford: Oxford University Press, 2015.

Senn, Matthias, ed. *Die Wickiana—Johann Jakob Wicks Nachrichtensammlung aus dem 16. Jahrhundert*. Zurich: Raggi, 1975.

Shapin, Steven. *The Scientific Revolution*. Chicago: University of Chicago Press, 1998.

Sirocko, Frank. *Wetter, Klima, Menschheitsentwicklung: von der Eiszeit bis ins 21. Jahrhundert.* 2nd ed. Darmstadt: WBG WissBuchges, 2010.

Slack, Paul. *The Invention of Improvement: Information and Material Progress in Seventeenth-Century England.* Oxford: Oxford University Press, 2014.

Slack, Paul, and Economic History Society. *The English Poor Law, 1531–1782.* Cambridge: Cambridge University Press, 1990.

Spinoza, Baruch de, ed. Wolfgang Bartuschat. Philosophische Bibliothek Band 95b: Baruch de Spinoza Sämtliche Werke Band 5b: *Politischer Traktat*: Lateinisch-Deutsch. 2nd edition, Meiner, F, 2010.

———. Sämtliche Werke / *Ethik in geometrischer Ordnung dargestellt*: Lateinisch-Deutsch. 2nd revised and improved edition, Meiner, F, 2007.

Spring, Eileen. *Law, Land, and Family: Aristocratic Inheritance in England, 1300 to 1800.* Durham: University of North Carolina Press, 2000.

Spufford, Margaret. *Small Books and Pleasant Histories: Popular Fiction and Its Readership in Seventeenth-Century England.* Cambridge: Cambridge University Press, 1985.

Squar, Michael. *Zwei Grad Celsius: Wie zügellose Gier die Welt verändert.* Munich: Buch & Media GmbH, 2015.

Stark, Rodney. *For the Glory of God: How Monotheism Led to Reformations, Science, Witch-Hunts, and the End of Slavery.* Princeton, NJ: Princeton University Press, 2003.

Stern, Philip J., and Carl Wennerlind. *Mercantilism Reimagined: Political Economy in Early Modern Britain and Its Empire.* New York: Oxford University Press, 2013.

Stone, Lawrence. *The Crisis of the Aristocracy: 1558–1641.* Oxford: Oxford University Press, 1965.

———. "The Educational Revolution in England, 1560–1640." *Past & Present*, 1964.

Talkenberger, Heike, ed. *Sintflut: Prophetie und Zeitgeschehen in Texten und Holzschnitten astrologischer Flugschriften 1488–1528.* Berlin and Boston: Walter de Gruyter, 1990.

Tate, William Edward. *The English Village Community and the Enclosure Movements.* London: Gollancz, 1967.

Taylor, Barry. *Society and Economy in Early Modern Europe, 1450–1789: A Bibliography of Post-War Research.* Manchester, UK: Manchester University Press, 1989.

Telelis, Ioannis. "The Climate of Tübingen A.D. 1596–1605 on the Basis of Martin Crusius' Diarium." *Environment and History* 4 (1998): 53–74.

Tenfelde, Klaus, and Hans Ulrich Wehler. *Wege zur Geschichte des Bürgertums: Vierzehn Beiträge.* Göttingen, Germany: Vandenhoeck & Ruprecht, 1994.

Therrell, Matthew, David Stahle, José Villanueva Diaz, Eliado Oviedo, and Malcolm Cleaveland, "Tree-Ring Reconstructed Maize Yield in Central Mexico: 1474–2001." *Climatic Change* 74 (2006): 493–504.

Tielhof, Milja van. *The "Mother of All Trades": The Baltic Grain Trade in Amsterdam from the Late 16th to the Early 19th Century.* Leiden: Brill, 2002.

Tinsley, Barbara Sher. *Pierre Bayle's Reformation: Conscience and Criticism on the Eve of the Enlightenment*. Selinsgrove and London: Susquehanna University Press and Associated University Presses, 2001.

Treml, Václav, Tereza Ponocná, Gregory M. King, and Ulf Büntgen. "A New Tree-Ring-Based Summer Temperature Reconstruction over the Last Three Centuries for East-Central Europe." *International Journal of Climatology*, November 2014.

Trepp, Anne-Charlott. *Im Zeichen der Krise: Religiosität im Europa des 17. Jahrhunderts*. Göttingen, Germany: Vandenhoeck & Ruprecht, 1999.

Tucker, Spencer C. *A Global Chronology of Conflict: From the Ancient World to the Modern Middle East*. Santa Barbara, CA: ABC-CLIO, 2009.

Tyacke, Nicholas. *Seventeenth-Century Oxford*. Oxford: Clarendon Press, 1997.

Utterström, Gustaf. "Climatic Fluctuations and Population Problems in Early Modern History," in *The Ends of the Earth*, translated by Donald Worster, pp. 39–79. New York: Cambridge University Press, 1988.

Vandenbossche, Hubert. *Adriaan Koerbagh en Spinoza*. Leiden: Brill, 1978.

van der Wal, Marijke J., and Gijsbert Rutten, eds. *Touching the Past: Studies in the Historical Sociolinguistics of Ego-Documents*. Amsterdam: John Benjamins Publishing, 2013.

van Deursen, A. Th. *Plain Lives in a Golden Age: Popular Culture, Religion and Society in Seventeenth-Century Holland*. Cambridge: Cambridge University Press, 1991.

Vermeer's Hat: The Seventeenth Century and the Dawn of the Global World. New York: Bloomsbury Press, 2008.

Vicente-Serrano, Sergio M., and José M. Cuadrat. "North Atlantic Oscillation Control of Droughts in North-East Spain: Evaluation since 1600 A.D." *Climatic Change* 85 (2007): 357–79.

Vittorio, Antonio Di. *An Economic History of Europe*. Abingdon, UK: Routledge, 2006.

Voges, Dietmar-Henning. *Nördlingen seit der Reformation: Aus dem Leben einer Stadt*. Munich: C. H. Beck, 1998.

Vollhardt, Friedrich. *Toleranzdiskurse in der Frühen Neuzeit*. Berlin and Boston: Walter de Gruyter, 2015.

Vries, Peer. *Ursprünge des modernen Wirtschaftswachstums: England, China und die Welt in der Frühen Neuzeit*. Göttingen, Germany: Vandenhoeck & Ruprecht, 2013.

Wallerstein, Immanuel. *The Modern World-System I: Capitalist Agriculture and the Origins of the European World-Economy in the Sixteenth Century*. Oakland: University of California Press, 2011.

———. *The Modern World-System II: Mercantilism and the Consolidation of the European World-Economy, 1600–1750*. Oakland: University of California Press, 2011.

Walter, John, and Roger Schofield. *Famine, Disease and the Social Order in Early Modern Society*. Cambridge: Cambridge University Press, 1991.

Warman, Arturo. *Corn and Capitalism: How a Botanical Bastard Grew to Global Dominance*. Durham: University of North Carolina Press, 2003.

Wefer, Gerold, Wolfgang H. Berger, Karl-Ernst Behre, and Eystein Jansen. *Cli-*

mate Development and History of the North Atlantic Realm. Springer Science & Business Media, 2013.

Weiguo, Liu, Liu Zhonghui, An Zhisheng, Wang Xulong, and Chang Hong. "Wet Climate During the 'Little Ice Age' in the Arid Tarim Basin, Northwestern China." *The Holocene* 21 (2011): 409–16.

Weikinn, Curt. *Quellentexte zur Witterungsgeschichte Europas von der Zeitwende bis zum Jahre 1850*. Berlin: Akademie-Verlag, 1958.

Wetter, O., and C. Pfister. "Spring-Summer Temperatures Reconstructed for Northern Switzerland and Southwestern Germany from Winter Rye Harvest Dates, 1454–1970." *Climate of the Past* 7 (2011): 1307–26.

White, Sam. "The Little Ice Age Crisis of the Ottoman Empire." In *Water on Sand*, herausgegeben von Alan Mikhail, 71–82. Oxford: Oxford University Press, 2012.

Whittington, Graeme. "The Little Ice Age and Scotland's Weather." *Scottish Geographical Magazine* 101 (1985): 174–78.

Wiesner-Hanks, Merry E. *Cambridge History of Europe. Vol. II: Early Modern Europe, 1450–1789*. 2nd ed. Cambridge: Cambridge University Press.

Wigley, T. M. L., M. J. Ingram, and G. Farmer. *Climate and History: Studies in Past Climates and Their Impact on Man*. Cambridge: CUP Archive, 1985.

Wilson, Cynthia. "The Little Ice Age on Eastern Hudson/James Bay: The Summer Weather and Climate at Great Whale, Fort George and Eastmain, 1814–1821, as Derived from Hudson's Bay Company Records." *Syllogeus* 55 (1985): 147–90.

Wilson, Peter H. *Europe's Tragedy: A New History of the Thirty Years War*. London: Penguin UK, 2009.

Wood, Andy. *Riot, Rebellion and Popular Politics in Early Modern England*. London: Palgrave Macmillan, 2001.

Wood, James B. "The Decline of the Nobility in Sixteenth and Early Seventeenth Century France: Myth or Reality?" *The Journal of Modern History* 48, no. 1 (1976): 1–29.

Woolf, D. R. *Reading History in Early Modern England*. Cambridge: Cambridge University Press, 2000.

Wootton, David. *The Invention of Science: A New History of the Scientific Revolution*. London: Allen Lane, 2015.

Wrightson, Keith. *English Society: 1580–1680*. New Brunswick, NJ: Rutgers University Press, 2003.

Xoplaki, Eleni, Panagiotis Maheras, and Juerg Luterbacher. "Variability of Climate in Meridional Balkans during the Periods 1675–1715 and 1780–1830 and Its Impact on Human Life." *Climatic Change* 48 (2001): 581–615.

Yadav, R. R., and J. Singh. "Tree-Ring-Based Spring Temperature Patterns over the Past Four Centuries in Western Himalaya." *Quaternary Research* 57 (2002): 299–305.

Zafirovski, Milan. *The Enlightenment and Its Effects on Modern Society*. Berlin: Springer Science & Business Media, 2010.

Zagorin, Perez. "The Crisis of the Aristocracy, 1558–1641, by Lawrence Stone." *The Journal of Economic History* 26, no. 1 (March 1966): 135–37.

Zittel, Claus. *Philosophies of Technology: Francis Bacon and His Contemporaries.* Leiden: Brill, 2008.

Zmora, Hillay. *Monarchy, Aristocracy and State in Europe 1300–1800.* Abingdon, UK: Routledge, 2002.

Zückert, Hartmut. *Allmende und Allmendaufhebung: Vergleichende Studien zum Spätmittelalter bis zu den Agrarreformen des 18./19. Jahrhunderts.* Stuttgart: Lucius & Lucius, 2003.

ILLUSTRATION CREDITS

Frontispiece and pages 6 and 7: Hendrick Averkamp: *Winter Landscape with Ice Skaters,* 1608, oil on canvas. Amsterdam, Rijksmuseum.

page 16: Pieter Bruegel the Elder, *The Harvesters (August),* 1565, oil on panel. New York, Metropolitan Museum of Art.

page 24: Pieter Bruegel the Elder, *Hunters in the Snow (Winter),* 1565, oil on panel. Kunsthistorisches Museum, Vienna, Gemäldegalerie.

page 28: Bird's Eye View of Amsterdam, woodcut, 1544. London, British Library.

page 33: Matthäus Merian the Elder, *Sinking of the Armada in 1588,* engraving. From: Johann Ludwig Gottfried, *Historische Chronica,* Frankfurt, 1630, p. 978. Berlin, Sammlung Archiv für Kunst und Geschichte.

page 36: The Great Fire of London, 1666, anonymous woodcut. Fototeca Gilardi.

page 41: Ciudad de Arequipa, 1613, woodcut. From: Huaman Poma de Ayala Felipe, *Nueva cronica y buen gobierno.* Biblioteca National Coleccion, Madrid.

page 47: Simon Bening (workshop), *Wine Harvest,* September, ca. 1513/15, tempera on vellum. From: Kalendarium Grimani, Venice, Biblioteca di San Marco.

page 59: Burning Witches on the Market Square in Guernesey, woodcut, ca. 1580.

page 70: Anonymous, *John Dee,* oil on wood, ca. 1594.

page 73: Johann Georg Mentzel, *Giordano Bruno,* engraving, 1548. Herzog August Bibliothek, Wolfenbüttel.

page 77: Anonymous, *Michel de Montaigne,* oil on wood, after 1581. Chantilly, Musée Condé.

page 86: Nicolas Robert, *Tulip Semper Augustus,* miniature, ca. 1675. Österreichische Nationalbibliothek, Vienna.

page 89: Theatrum anatomicum at Leiden University, anonymous engraving, ca. 1610.

page 94: Rembrandt van Rijn, *Self Portrait with Cap and Opened Mouth*, etching, 1630. Liszt Collection.

page 111: Musical Instruments, woodcut, 1654. From: Johann Amos Comenius, *Orbis sensualium.*

pages 124–25: Matthäus Merian, *The Battle of Nördlingen, 6 September 1634*, engraving. From: M. Merian, *Theatrum Europaeum*, 1670. Bayerische Staatsbibliothek, Munich.

page 127: Exercising with a Musket, engraving. From: Jacques de Gheyn, *Waffenhandlung von den Rören, Musquetten undt Spiessen*, Gravenhagen, 1608. Bayerische Staatsbibliothek, München.

page 140: African Slaves Producing Sugar for Export in the Spanish Colonies in America, engraving. From: Theodor de Bry, *America pars quinta*, Frankfurt, 1594.

page 143: Jacob Waben, *Portrait of Jan Pieterszoon Coen*, oil on canvas, early seventeenth century.

page 156: Anonymous, *Christina of Sweden, Elisabeth of Bavaria and René Descartes*, oil on canvas. Fototeca Gilardi.

page 159: Claude Duflos the Elder, *Marin Mersenne*, engraving after a contemporary portrait. Paris, private collection.

page 166: Robert Nanteuil, *Pierre Gassendi*, portrait, engraving, 1658. Berlin, Sammlung Archiv für Kunst und Geschichte.

page 171: Section of the Earth with Central Fire, Other Fiery Centers and Volcanos, anonymous engraving. From: Athanasius Kircher, *Mundus subterranus*, Amsterdam, 1664.

page 176: Hans Tisseur, *The Rasphuis in Amsterdam*, engraving. From: Melchior Fokkens, *Beschrijvinge der wildt-vermaardt Koop-Stadt Amstelredam*, 1662. Leiden, University Library.

page 179: Beached Whale on the Dutch Coast Close to Scheveningen in February 1598, contemporary engraving.

pages 182 and 184: Maria van Oosterwijck, *Vanitas Still Life*, 1668, oil on canvas. Kunsthistorisches Museum Wien, Gemäldegalerie.

page 187: Johann Jacob Schollenberger, *Comet above the City of Nuremberg in November and December 1680*, engraving, 1680.

page 189: Pierre Bayle, engraving after a portrait, 1675.

page 197: Pierre Bayle, *Dictionnaire historique et critique*, page with the beginning of the article on Epicurus. Volume II, p. 364. Bayerische Staatsbibliothek Munich.

page 201: Baruch de Spinoza, anonymous contemporary etching after a portrait, ca. 1665.

page 222: Schabbatai Zevi, contemporary engraving. Bildarchiv Pisarek.

page 227: Frost Fair on the River Thames in London, 1683, anonymous woodcut.

page 234: Portrait of John Locke, engraving by George Vertue after a portrait by Godfrey Kneller, 1697.

page 239: Louis XIV as Apollo/Rising Sun in the Ballet Royal de la Nuit, 23 February 1653, watercolour. Costume design by Henri de Gissey. Paris, Bibliothéque Nationale.

page 252: Henry Raeburn, *The Reverend Robert Walker Skating on Duddingston Loch*, oil on canvas, 1790s. The National Galleries of Scotland, Edinburgh.

INDEX